Liquid Institutionalization at Sea

Jan P. M. van Tatenhove

Liquid Institutionalization at Sea

Reflexivity and Power Dynamics of Blue
Governance Arrangements

Jan P. M. van Tatenhove
Centre for Blue Governance
Aalborg University
Aalborg, Denmark

ISBN 978-3-031-09770-6   ISBN 978-3-031-09771-3   (eBook)
https://doi.org/10.1007/978-3-031-09771-3

This Palgrave Macmillan imprint is published by the registered company Springer Nature Switzerland AG.
The registered company address is: Gewerbestrasse 11, 6330 Cham, Switzerland

PREFACE

This book is developed from the idea to give an inaugural lecture about liquid institutionalization for taking up the post of professor of Marine Governance and Maritime Spatial Planning at the Centre for Blue Governance (CBG), Department of Planning at Aalborg University (AAU) in Denmark. The inaugural lecture was planned on 26 May 2020 as closure of the festivities related to the opening of the Centre for Blue Governance. This date was chosen to connect to the final meeting of the COST Action "Ocean Governance for Sustainability—challenges, options and the role of science" (OceanGov) which was organized by CBG in cooperation with ZMT Leibniz-Zentrum für Marine Tropenforschung in Bremen (as the overall coordinator of this COST action) and should take place from 27–29 May 2020.

However, the Covid-19 pandemic and the restrictions from March 2020 changed all these plans. Instead of a festive opening of the Centre for Blue Governance and the celebration of the successful COST Action OceanGov project, Europe, Denmark and Aalborg University locked down. Meetings, teaching activities and conferences became on-line events. After one and a half year, and three consecutive periods of closing down societies and short periods of opening up, in November 2021 countries all over the world are facing the fourth period of closing down. Although in autumn 2021 many European people were vaccinated, infections rose again and the upcoming of new Covid-19 variants, such as Delta and Omicron,[1] resulted in a fourth period of Covid-19 restrictive measures and lockdowns. In the spring of 2022, societies reopened and it was possible to have physical meetings again; a "new normal" institutionalized!

Initially I planned to publish my CBG inaugural lecture as a booklet and/or article as I did with my inaugural lectures at Wageningen University (27 October 2011) and Queens's University Belfast (27 April 2016). I did not have the intention to write a book, but this changed when I was approached by Rachel Ballard of Palgrave. She sent me a mail the beginning of May 2020 to have a phone/Skype call about a possible book project. After our talk on the 12th of May the idea to write a book, in the Palgrave Pivot series matured in my head and in July 2020, I decided to develop my ideas for the inaugural lecture into this book project.

All my research projects and teaching activities have been collaborative endeavours, in which I cooperated with colleagues from different departments, sections and universities nationally and internationally (Nijmegen, Amsterdam, Wageningen, Belfast, Aalborg). Due to the specific circumstance in 2020 and 2021 the writing of this book has been mainly a solitary activity, but builds upon my earlier research projects, teaching activities and other projects I have been involved in in the last 25 years with a focus on processes of institutionalization and the development of governance/policy arrangements in environmental policy, nature conservation, spatial planning and marine policy domains.

This book is not the product of Covid-19 but was only possible due to the Covid-19 pandemic. Despite the difficult and tough times of social isolation (living in Denmark and physically removed from family and friends in the Netherlands), this period gave me the extra time in the evenings and weekends to do the writing. Also the paradigmatic shift from physical, face-to-face teaching activities, presentations, workshops, working groups, conferences and congresses to on-line meetings saved a lot of time, because time-consuming travels were not possible.

Because this book is the result of my research of the last 25 years, it is indirectly the product of cooperation and collaboration with a lot of colleagues and friends. Starting in the "here-and-now" I like to thank the colleagues of the Centre for Blue Governance: Jesper Raakjær, Alyne

---

[1] On 26 November 2021, WHO designated the variant B.1.1.529 a variant of concern, named Omicron, on the advice of WHO's Technical Advisory Group on Virus Evolution (TAG-VE) (https://www.who.int/news/item/28-11-2021-update-on-omicron, visited 01/12/2021).

Delaney, Troels Jacob Hegland, Søren Qvist Eliasen, Rikke Becker Jacobsen, Kristen Ounanian, Paulina Ramírez Monsalve, Mette Møller Jeppesen, Päivi Haapasaari, Mathilde Højrup, Trine Skovgaard Kirkfeldt, Holly Niner, Nelson F. Coelho, Matthew Howells, Juliane Kirstine Bjerring, Sun Cole Seeberg Dyremose and Birgitte Bauer. I have to admit the preparation of this inaugural lecture of 45 minutes got out of hand, resulting in this book. Second the former colleagues of the Public Policy Group of the University of Amsterdam: Maarten Hajer, John Grin, Liesbeth Bervoets and Anne Loeber. With them I broadened my approach with insights from Multi-Level Governance, deliberative and participatory democracy, interactive policy-making and power in organizations. Third, the former colleagues of the Environmental Policy Group at Wageningen University, the Netherlands who applied and further developed the Policy Arrangements Approach in their PhD or other research projects: Judith van Leeuwen, Luc van Hoof, Birgit de Vos, Hilde Toonen, Chris Seijger, Dorien Korbee, Stephanie Janssen, Linde van Bets, Harry Barnes-Dabban, Coco Smits and Eira Carballo Cardenas. Last, but not least I like to thank the former colleagues of the MNL (*Milieu, Natuur* and *Landschap*) group (later Department of Environmental Policy Sciences) at the Radboud University Nijmegen (the Netherlands). Together with Bas Arts and Pieter Leroy I developed the Policy Arrangement Approach (PAA), and with Marielle van der Zouwen, Froukje Boonstra and Duncan Liefferink we further developed the PAA. With Paul Pestman I developed my first ideas about reflexivity, and with Bas Arts the multi-layered framework of power.

I hope you enjoy reading this book about liquid institutionalization.

Nørresundby, Denmark; Leeuwarden,   Jan P. M. van Tatenhove
The Netherlands
March 2022

CONTENTS

LIST OF FIGURES

CHAPTER 1

Introduction

Abstract This introductory chapter gives an overview of different conceptualizations of institutionalization to understand the dynamics of stability and change in policy-making processes at sea. Institutionalization is on the one hand about processes in which norms, values, roles, perceptions and ways of interaction are created and stabilized in more or less fixed patterns of rules, organizational structures and institutions. On the other, it is about the decay of institutions and/or processes in which stabilized and fixed rules of the game, organizational structures and institutions are challenged. The chapter starts with an overview of how institutionalization is defined in the social sciences. This book contributes to the debates on institutionalization by introducing and analyzing the core concept of liquid institutionalization to understand processes of change and stability of blue governance arrangements at seas. In Sect. 1.3 the structure of the book is presented.

Keywords Institutionalization • Liquid institutionalization • Change • Stability

1.1   INTRODUCTION

This book is about understanding (institutional) change and stability in policy-making and governance processes at sea. More specifically, it seeks to understand the enabling and constraining conditions to govern maritime activities within specific institutional contexts. Intriguing questions for social and political scientists are what social and political processes are taking place at seas, who is responsible, who has the authority to take decisions, who is included or excluded in these processes of governance, and how do processes of institutionalization enable and constrain governance at seas?

Not many people live at sea, but many are involved in maritime activities such as fishing, navigation/shipping, extracting minerals and natural resources (oil and gas production, deep seabed mining, navigational dredging), developing or working on offshore wind parks and enjoying recreation at beaches and on shorelines. While these maritime activities contribute substantially to the GDP (Gross Domestic Product) of (inter) national economies, they are coupled with severe environmental, spatial and social problems, such as the overexploitation of fish stocks, the depletion of natural resources, the emissions of Greenhouse Gases (GHG), the damage to and threatening of marine ecosystems by oil spills, excessive discharges of chemicals, toxic substances, garbage, invasive non-indigenous species and the peripheralization of coastal communities.

However, governance at seas is a policy paradox (Stone 2012). While there is an urgent need to find solutions for a diversity of environmental problems, including climate change, and spatial conflicts, the governing of these problems is very complicated, because there is not one sole authority responsible for these problems, nor do these problems stop at the political and administrative borders of states. Clear governance and planning structures are missing, while much of the maritime activities and their consequences are invisible not only for ordinary people, but also for authorities. When moving further away from the shore authority, territory and rights (Sassen 2006) changes, which will affect the possibilities to govern maritime activities and their consequences in a legitimate way beyond the territorial waters of states at regional seas and high seas/Areas Beyond National Jurisdiction (ABNJ).

Policy-making and governance processes do not take place in a vacuum, but happen in the context of pre-existing institutions and organizations which influence the interactions and negotiations which can develop between governmental and non-governmental actors. In the context of this book, these enabling and constraining conditions of policy-making and governance processes are understood as processes of institutionalization.

1.2   INSTITUTIONALIZATION

Institutionalization is a core concept in sociology, organization theory, political science and public administration, used to understand the dynamics of stability and change in organizations, political systems and societies in general. On the one hand it is about processes in which norms, values, roles, perceptions and ways of interaction are created and stabilized in more or less fixed patterns of rules, organizational structures and institutions. On the other, institutionalization is about the decay of institutions and/or processes in which stabilized and fixed rules of the game, organizational structures and institutions are challenged, resulting in a power game to change, adapt or redirect existing organizational structures and institutions or to develop new ones.

In general institutionalization is viewed as a social process in which actors accept a shared definition of reality (Scott 1987). According to Scott (1987: 496), actors take this conception "for granted as defining the 'way things are' and/or the 'way things are to be done'". Also (Selznick 1957) sees institutionalization as a historic process of instilling values, stressing stability and the persistence of structure over time. A very influential definition is presented by Berger and Luckmann in 1966, stating that processes of habitualization (i.e. action that becomes cast into patterns and may be performed in the future in the same manner) preceded any institutionalization (Berger and Luckmann 1991: 71). Institutionalization occurs whenever there is a reciprocal typification of habitualized actions by types of actors. Put differently, any such typification is an institution" (Berger and Luckmann 1991: 72). For reciprocal typification to occur there must be a continuing social situation in which the habitualized actions of two or more individuals interlock (1991: 75). When a third actor enters, the stage interactions between A and B will change and the existing institutional world of A and B is now passed to others. In the words of Berger and Luckmann (1991: 76) "the habitualizations and typifications undertaken in the common life of *A* and *B*, formations that until this point still had the quality of *ad hoc* conceptions of two individuals, now become historical institutions" (1991: 76). This institutional world is experienced by actors as an objective reality; "institutions *are* there, external to him, persistent in their reality, whether he likes it or not" (78). However, according to Berger and Luckmann "(...) the objectivity of the institutional world, however massive it may appear to the individual, is a humanly produced, constructed objectivity. (...) The institutional world is objectivated human activity, and so is every single institution." This results in a paradox: actors can produce the social world, which the actor thereafter experiences as something other than a human product. This is what Berger and Luckmann call a dialectic relationship. "Externalization and objectivation are moments in a continuing dialectical process" (78). The third moment in this process is "internalization, by which the objectivated social world is retrojected into consciousness in the course of socialization" (78–79). In a nutshell institutionalization is about the forming of institutions (social world) by actors in interactions, these institutions are perceived as an objective reality affecting the types of interactions which can take place, but institutionalization is not an irreversible process (institutions can change in human interactions).

Initially institutionalization was primarily studied as an interorganizational process. Guy Peters and Pierre (Guy Peters and Pierre 2012) cite the definition of institutionalization of Selznick "infusing values into structure" (pp. viii), which suggests that creating an institution is an internal organizational only, neglecting the extra-organizational aspects (Pierre and Guy Peters 2009). The concept is applied in several social sciences and domains to understand processes of change and stability. For example Zucker (Zucker 1977) builds upon the conception of Berger and Luckmann. She defines institutionalization as "both a processes and a property variable. It is the process by which individual actors transmit what is socially defined as real, at the same time, at any point in the process the meaning of an act can be defined as more or less a taken-for-granted part of this social reality. Institutionalization acts, then, must be perceived as both objective and exterior" (Zucker 1977: 728). Zucker (Zucker 1977) studied the effect of different degrees of institutionalization in constructed realities on cultural persistence in three experiments (transmission, maintenance and resistance to change). She concluded that the greater the degree of institutionalization, the greater the generational uniformity of cultural understandings, the greater the resistance to change.

According to Pierre and Guy Peters (2009), until recently, institutionalization was underdeveloped in public administration and political science, especially the relation of institutions in the context of their environment and the question of agency. They distinguish two conceptualizations of institutionalization: first, a sociological organization theory in which an institution involves a normative commitment to the structure and its role in society, and secondly, a structural-procedural theory which requires building formal structures and procedures or the routinization of informal practices to enhance the predictability and to carry out assigned tasks.

Overall, in sociological and political science approaches the importance of internal factors in institutionalization is emphasized. For example Huntington (cited in Blondel 2008: 717) defines institutionalization in 1968 as "the process by which organisations and procedures acquire value and stability." However, also the environment (consisting of other institutions and organizations) and the way actors perceive and interpret other institutions is crucial to understand processes of institutionalization. The environment may have objective features but is at the same time a social construction in which actors interpret the environment and its relation with the single institution. Actors are capable of breaking down patterns of behaviour or even existing institutions and questioning existing institutionalized values.

There are several examples of institutionalization, such as models of parties and party institutionalization by (Randall and Svåsand 2002), and the analysis of institutional isomorphism to compare institutionalization in public and private organizations (Frumkin and Galaskiewicz 2004). From planning theory Alexander (Alexander 2005) approached institutionalization from a planning design approach, conceptualizing institutionalization as the result of conscious choices, instead of a natural social process. Institutional design means designing institutions; rules, procedures and organizational structures enabling and constraining agency. According to Alexander the development of (human) institutions is the result of intentional decisions. He developed a multi-level model of institutional design (ID). On the macro level ID is applied to societies or addresses significant macro-societal processes and institutions. It concerns the development and design of constitutional rules, political administrative programmes etc. by statesmen and politicians. On the meso-level, ID is about planning and the implementation of structures and processes, coordinated by professional planners. The micro-level of ID involves intra-organizational design of formal-informal units, processes and interactions, such as committees, teams and task forces.

In general institutionalization is an ongoing process of patterning, preservation, construction, organization and deconstruction of day-to-day activities and interactions in institutions (van Tatenhove and Leroy 2000; van Tatenhove 1993). To understand the dynamics of governance and policy-making it is important to understand the transformation processes in which institutions and institutional rules are produced, formed, defended and reproduced by actors in interactions. These interactions do not take place in a vacuum but in an historically formed (institutional) environment, which enables and constrains the interactions between actors and the way problems and solutions are defined.

This book will contribute to the debate on institutionalization by introducing the concept of liquid institutionalization, to understand processes of change and stability of governance arrangements at sea. In sociology, economics, and political theory institutionalization refers to the process in which fluid behaviour gradually solidifies into institutional rules (van Tatenhove and Leroy 2000). In this process of formation of institutional rules (process of structuration), institutions are formed. In this book, the focus will be on the institutionalization of governance arrangements at sea, more specifically the way the structural properties of these blue governance arrangements (the rules and resources) are formed in the interaction between a diversity of actors. Once governance arrangements have been produced (process of stabilization) they are defended against change. In contrast to "normal" processes of institutionalization, liquid institutionalization emphasizes continuous change. Key concepts of this theory of liquid institutionalization are blue governance arrangements, institutional change, power and reflexivity. I will apply this theory of liquid institutionalization on three different cases: Arctic Shipping, Deep Seabed Mining and Transboundary Regionalization in Europe.

Neo-institutional theories focus on stability and not so much on change (Mahoney and Thelen 2010). To understand processes of institutionalization at sea, the main argument of this book is that it is crucial to understand process of change from both a structural and an agency perspective. In situations of ambiguity, it could be that rules are continuously negotiated and are in a permanent state of "flux and becoming," sometimes without a solid stabilization in institutionalized governance arrangements. This book will analyse how blue governance arrangements in the three maritime domains institutionalize and whether one can speak of processes of continuous structuration, in other words liquid institutionalization.

1.3   STRUCTURE OF THE BOOK

Chapter 2 presents the theory of liquid institutionalization at sea. The main concepts of this theory are institutionalization, blue governance arrangements, institutional change, power and reflexivity. Liquid institutionalization is based on a liquid ontology. That is, the study of the nature of social being at sea, the kind of things that exist in the reality at sea, the conditions of existence, the relations of dependency and the standards that have to be met for something to fully exist at sea (Elder-Vass 2013; Kauppi 2018; Scott and Marshall 2009). I define liquid institutionalization as an ongoing process of structuration in which blue governance arrangements are continuously negotiated and in a permanent state of "flux and becoming" without a solid stabilization in institutionalized blue governance arrangements. This reflects high institutional ambiguity (van Leeuwen et al. 2012). A blue governance arrangement is the temporary stabilization of the substance and organization of a marine policy domain, in terms of coalitions of actors, discourses, resources and rules of the game (van Tatenhove et al. 2000). Power is the "organisational and discursive capacity of agencies, either in competition with one another or jointly, to achieve outcomes in social practices, a capacity which is however co-determined by the structural power of those social institutions in which these agencies are embedded" (Arts and van Tatenhove 2005: 347). Power is a multi-layered phenomenon, consisting of relational, dispositional and structural relational power. Reflexivity refers to the capacity of actors to govern and to induce institutional change (i.e. to change the processes of structuration and stabilization) by challenging the existing discursive spaces of marine governance arrangements (performative mobilization), and to activate and to use rules and resources from different rule systems and layers of government (van Tatenhove 2017). The conceptual framework consists of a continuum of types of liquid institutionalization, from frozen to liquid, expressing different forms of institutional change, reflexivity and specific forms of power.

Chapters 3, 4, and 5 present the (liquid) institutionalization of different maritime activities. Chapter 3 discusses the possible future development of Arctic shipping. The chapter opens with three possible future Arctic shipping routes (the North-West Passage (NWP); the Northeast Passage and Northern Sea Route (NEP/NSR); and the Transpolar Sea Route (TSR)) and types of Arctic shipping, such as liner shipping, bulk shipping (liquid and dry), specialized shipping (LNG and reefer) and cruise shipping. For each of these shipping routes the existing and emerging shipping governance arrangements will be constructed. The chapter ends with the specific forms of liquid institutionalization of Arctic shipping and related forms of power and reflexivity. Chapter 4 analyses the developments within Deep Seabed Mining (DSM). The chapter starts with discussing the governance context and practice of DSM, followed by the role of the International Seabed Authority, and the debate about the pros and cons of the exploration and exploitation of marine resources from the seabed and the way this is organized. This is illustrated with the example of the *Clarion Clipperton Zone*, in the Pacific Ocean. Based on this description, emerging deep seabed mining governance arrangements are constructed and their specific form of liquid institutionalization. Chapter 5 deals with forms of transboundary maritime regionalization in Europe. Examples are Transboundary Maritime Spatial Planning (e.g. in the North Sea, the Baltic Sea and in the Atlantic Ocean), the development of Sea Basin Strategies (e.g. for the Black Sea and the Arctic Ocean), and Marco-Regional Strategies (for the Baltic Sea Region and for the Adriatic and Ionian Region). For each of these examples of transboundary maritime regionalization, emerging transboundary governance arrangements are constructed with their specific forms of liquid institutionalization. Chapter 6 presents the conclusions and reflections. After summarizing the main arguments and findings, this chapter gives insights in what we can learn from the different cases. The focus will be on forms of reflexivity and power in different forms of liquid institutionalization and the consequences (enabling and constraining conditions) for designing and developing effective and legitimate blue governance arrangements. The chapter ends by formulating future research needs.

References

Alexander, E.R. 2005. Institutional Transformation and Planning: From Institutionalization Theory to Institutional Design. *Planning Theory* 4 (3): 209–223. https://doi.org/10.1177/1473095205058494.

Arts, Bas, and Jan P.M. van Tatenhove. 2005. Policy and Power: A Conceptual Framework Between the 'Old' and 'New' Policy Idioms. *Policy Sciences* 37 (3–4): 339–356. https://doi.org/10.1007/s11077-005-0156-9.

Berger, Peter, and Thomas Luckmann. 1991. *The Social Construction of Reality. A Treatise in the Sociology of Knowledge*. London: Penguin Books Ltd. Reprint.

Blondel, Jean. 2008. *About Institutions, Mainly, but Not Exclusively, Political.* Oxford: Oxford University Press. https://doi.org/10.1093/oxfordhb/9780199548460.003.0036.

Elder-Vass, Dave. 2013. *The Reality of Social Construction.* Cambridge, UK: Cambridge University Press.

Frumkin, P., and Joseph Galaskiewicz. 2004. Institutional Isomorphism and Public Sector Organizations. *Journal of Public Administration Research and Theory* 14 (3): 283–307. https://doi.org/10.1093/jopart/muh028.

Guy Peters, B., and Jon Pierre. 2012. Introduction: New Research Agenda. In *Institutionalism II*, ed. B. Guy Peters and Jon Pierre, vii–xi. London: SAGE Publications Ltd.

Kauppi, Niilo. 2018. *Toward a Reflexive Political Sociology of the European Union. Fields, Intellectuals and Politicians.* Palgrave Macmillan UK. https://doi.org/10.1007/978-3-319-71002-0.

Mahoney, James, and Kathleen Thelen. 2010. A Theory of Gradual Institutional Change. *Explaining Institutional Change: Ambiguity, Agency, and Power.* 1–37. https://doi.org/10.1017/9780521134323.

Pierre, Jon, and B. Guy Peters. 2009. From a Club to a Bureaucracy: JAA, EASA, and European Aviation Regulation. *Journal of European Public Policy* 16 (3): 337–355. https://doi.org/10.1080/13501760802662706.

Randall, Vicky, and Lars Svåsand. 2002. Party Institutionalization in New Democracies. *Party Politics* 8 (1): 5–29. https://doi.org/10.1177/1354068802008001001.

Sassen, Saskia. 2006. *Territory, Authority, Rights. From Medieval to Global Assemblages.* Princeton and Oxford: Princeton University Press.

Scott, W. Richard. 1987. The Adolescence of Institutional Theory. *Administrative Science Quarterly* 32 (4): 493–511. https://doi.org/10.2307/2392880.

Scott, John, and Gordon Marshall. 2009. *A Dictionary of Sociology.* 3rd ed. Oxford, UK: Oxford University Press. https://doi.org/10.1093/acref/9780199533008.001.0001.

Selznick, P. 1957. *Leadership in Administration.* New York: Harper and Row.

Stone, Deborah. 2012. *Policy Paradox: The Art of Political Decision Making.* 3rd ed. New York, NY: W.W. Norton & Company, Inc.

van Leeuwen, Judith, Luc van Hoof, and Jan P.M. van Tatenhove. 2012. Institutional Ambiguity in Implementing the European Union Marine Strategy Framework Directive. *Marine Policy* 36 (3): 636–643. https://doi.org/10.1016/j.marpol.2011.10.007.

van Tatenhove, Jan P.M. 1993. *Milieubeleid Onder Dak? Beleidsvoeringsprocessen in Het Nederlandse Milieubeleid in de Periode 1970–1990, Nader Uitgewerkt Voor de Gelderse Vallei.* Wageningen: Pudoc.

————. 2017. Transboundary Marine Spatial Planning: A Reflexive Marine Governance Experiment? *Journal of Environmental Policy and Planning* 19 (6): 783–794. https://doi.org/10.1080/1523908X.2017.1292120.

van Tatenhove, Jan P.M., and Pieter Leroy. 2000. "The Institutionalisation of Environmental Politics." In *Political Modernisation and the Environment. The Renewal of Environmental Policy Arrangements*, edited by Jan P.M. van Tatenhove, Bas Arts, and Pieter Leroy, 17–33. Dordracht/Boston/London: Kluwer Academic Publishers.

van Tatenhove, Jan P.M., Bas Arts, and Pieter Leroy. 2000. *Political Modernisation and the Environment. The Renewal of Environmental Policy Arrangements*. Ed. Jan P.M. van Tatenhove, Bas Arts, and Pieter Leroy. Dordracht/Boston/ London: Kluwer Academic Publishers.

Zucker, Lynne G. 1977. The Role of Institutionalization in Cultural Persistence. *American Sociological Review* 42: 726–743.

CHAPTER 2

Liquid Institutionalization

Abstract This chapter presents the analytical framework of liquid institutionalization. This analytical framework consists of five analytical concepts: liquid institutionalization, blue governance arrangements, institutional change, reflexivity and power. Liquid institutionalization is defined as an ongoing process of structuration/elaboration in which blue governance arrangements are continuously negotiated, while in a permanent state of "flux and becoming," as a consequence of high ambiguity and reflexivity, without a solid stabilization in institutionalized blue governance arrangements. Characteristic of liquid institutionalization is its liquid ontology, a detailed understanding of institutional change and stability and how governance arrangements institutionalize, due to the interplay of political modernization and interactions in policy practices. As an ideal-type liquid institutionalization emphasizes continuous processes of structuration. Its ambiguity makes it possible to continuously negotiate the rules of the game. To understand this permanent state of negotiation and change, liquid institutionalization is conceptualized as a form of reflexivity and power. The chapter concludes with a continuum of liquid institutionalization, consisting of four ideal types based on the building blocks elaboration/ structuration—maintenance/stabilization, forms of institutional change and forms of reflectiveness/reflexivity.

Keywords Liquid institutionalization; Blue governance arrangements; Liquid ontology • Institutional change • Reflexivity • Power

2.1   INTRODUCTION

This chapter presents the theoretical framework of the book: the liquid institutionalization of blue governance arrangements. The chapter starts with defining blue governance arrangements (Sect. 2.1), followed by liquid institutionalization (Sect. 2.2). To understand the dynamics of processes of liquid institutionalization, the Sects. 2.3 and 2.4 discuss forms of reflexivity and power. The chapter ends with presenting a continuum of liquid institutionalization (Sect. 2.5).

2.2   BLUE GOVERNANCE ARRANGEMENTS

Marine governance is "the sharing of policy making competencies in a system of negotiation between nested governmental institutions at several levels (international, supranational, national, regional and local) on the one hand, and state actors, market parties and civil society organizations on the other hand in order to govern activities at sea and their consequences" (van Leeuwen and van Tatenhove 2010: 591). This definition of marine governance emphasizes the joint competences of public and private actors—in multi-level governance settings—to govern activities. However, this definition does not give insight into the capacity of actors to anticipate contemporary and future developments, nor their ability to change discursive spaces and institutional rules. To understand anticipatory capacity building processes and the reflexive capacity of public and private actors to define, address and change developments and processes related to resource exploitation at seas in processes of institutionalization, I will use the concept of blue governance instead of marine governance. Marine governance focuses on seas and oceans, more specific marine ecosystem processes and maritime activities at seas, while blue governance also addresses the dynamics of deltas and land-sea interactions. Blue governance is the capacity of public and private actors within blue governance arrangements to govern maritime activities, including land-sea interactions and activities in the deltas, and their consequences in anticipatory and reflexive ways. Governance processes continuously change, stabilize and change again. To analyse and explain these processes of stability and change I understand blue governing as a process of liquid institutionalization in which public and private actors in interactions in a reflexive and anticipatory way produce and reproduce, defend and conquer the structural conditions (rules, resources and discourses) of blue governance arrangements within fragmented institutional settings at sea.

In general, governance arrangement refers to the way a policy domain is temporarily shaped in terms of substance and organization (Liefferink 2006; van Tatenhove 2013). Substance refers to discourses, resulting in distinct policy and regulatory goals and objectives, whereas organization refers to the coalitions established by the actors involved, the rules of the game (instruments, procedures, division of tasks) and the available resources. In a blue governance arrangement, *coalitions* of governmental and non-governmental maritime actors try to influence activities and developments (in an anticipatory and reflexive way) within a specific maritime domain or sector. In interactions these actors negotiate, develop and design legitimate initiatives, institutions and solutions, based on specific *discourses*, the ability to mobilize and to use *resources* and to define the *rules* of the game on different levels (based on van Tatenhove 2013: 298). In other words, the structure of a blue governance arrangement can be analysed along four dimensions: actors and their coalitions, resources, rules of the game, and discourses (van Tatenhove et al. 2000; Liefferink 2006). *Actors and their coalitions* are public and private actors, organization and agencies involved in the development of marine politics and policies. *Resources* refer to the unequal division of resources among these actors, which leads to differences in power and influence. Examples of resources are money, information, permits, knowledge or expertise. *Rules of the game* in marine policies and politics refer to the formal rules and procedures in the different stages of the policy-making process (agenda setting, policy formulation, decision-making, implementation, enforcement and evaluation) and the informal rules and "routines" of interaction within marine practices and the relevant institutions of marine politics and policy-making. *Discourses* entail the norms and values, as well as the definitions of problems and approaches to solutions by the actors involved. A discourse is the specific ensemble of ideas, concepts and categorizations through which meaning is given to physical and social realities (Hajer 1995). Marine and maritime discourses refer to ideas about the character and definitions of problems related to marine ecosystems and/or defined by maritime sectors, their causes and perceived solutions.

In blue governance arrangements, actors not only have the capacity to govern maritime activities, but also are able to reflect upon the consequences of their actions and activities for both communities, maritime sectors and marine ecosystems in the past, the present and in the future.

Therefore, to understand the dynamics of maritime policy-making and politics, it is crucial to understand the governance dynamics of different maritime sectors (both for each sector separately and in interactions with each other) and to have insight in the specific institutionalization of blue governance arrangements; how do blue governance arrangements change and how are they defended against change? This process of institutionalization of blue governance arrangements refers to the ordering of a specific maritime policy field (such as shipping, deep seabed mining, fisheries, aquaculture, offshore renewable energy production, etc.) in terms of actors/coalitions, resources, rules and discourses. The changes in the dimensions of a governance arrangement are the result of processes of political modernization on the one hand, that is, the interplay of contextual processes of structural political and social change, and problem-oriented renewal of policy-making and decision-making by agents in day-to-day (policy) practices on the other (see Fig. 2.1).

2.3   LIQUID INSTITUTIONALIZATION

To understand processes of institutionalization at sea I introduce the concept of liquid institutionalization. Why liquid? What is the added value? An important justification for understanding institutionalization processes at sea as liquid has to do with the characteristics of governance processes at sea; they are mainly invisible for citizens, there is not one responsible authority, there are no property rights and sectors have their own rules and governance systems. Liquidity at sea does not in this case refer to the

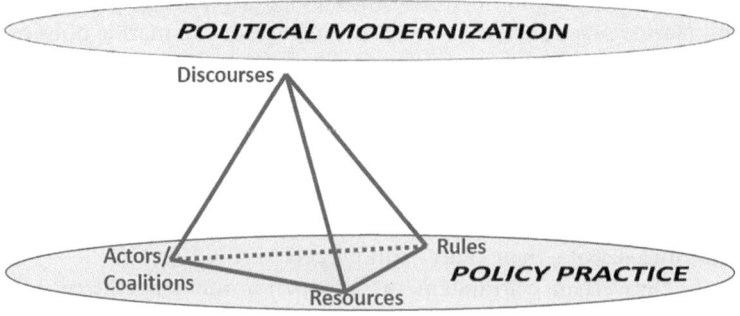

Fig. 2.1   The interplay between political modernization, policy practice and governance arrangements

water (waves and currents), nor the liquidity of activities (navigation) and resources (fish), but to the inherent flux and the permanent state of becoming of governance rules and governing institutions at sea.

Understanding liquid institutionalization requires a thorough understanding of institutionalization as a process (Sect. 2.3.1). Liquid institutionalization is an ongoing process of structuration in which blue governance arrangements are continuously negotiated, while in a permanent state of "flux and becoming" without a solid stabilization in institutionalized blue governance arrangements. This process of liquidity is the result of the fragmented setting beyond territorial seas, at the regional seas and the high seas. Characteristic for this fragmented institutional setting at sea is a state of institutional ambiguity (van Leeuwen et al. 2012), which refers to a situation of mismatch between institutions and institutional rules from different institutional settings (e.g. Regional Sea Commissions, European Union (EU), and United Nations (UN) institutions). This gives actors room to manoeuvre and to select and use rules from these different (governmental) levels when negotiating new institutional rules to replace existing ones. In this sense, liquidity as introduced in this chapter, is a form of ambiguity, for example the possibility to continuously negotiate the rules of the game, which could increase the reflexivity of actors. In the next sub-sections I will discuss institutionalization (Sect. 2.3.1), its liquid ontology (Sect. 2.3.2) and how to understand processes of change and stability in processes of liquid institutionalization (Sect.2.3.3).

2.3.1 What Is Institutionalization?

Institutionalization is the process of production and reproduction of governance arrangements, in which public and private actors produce and reproduce the rules and resources in interaction within the context of long-term processes of societal and political transformation, conceptualized as political modernization (van Tatenhove 1998; Arts et al. 2006; Arts and van Tatenhove 2006; Liefferink 2006; van Tatenhove et al. 2000). Arts and van Tatenhove (2006) distinguished different conceptualizations of political modernization. A normative one in day-to-day politics, referring to needed institutional reforms, an analytical conceptualization to understand shifts in governance and an analytical political sociological conceptualization to understand the effects of structural processes on day-to-day policy-making by focusing on new relationships between state, civil society and market (van Tatenhove 1999). They define political modernization "as the shifting relationships between the state, market and civil society in political domains of societies—*within* countries and *beyond*—as a manifestation of the 'second stage of modernity', implying new conceptions and structures of governance" (Arts and van Tatenhove 2006: 29).

Institutionalization is about change and stability, which is reflected in two sub-processes: structuration and stabilization (van Tatenhove 1993; van Tatenhove and Leroy 2000; van Tatenhove et al. 2021). Structuration is the gradual formation and production of structural properties of a governance arrangement in interaction. Specific forms of interaction between public and private actors (forming coalitions) within relations of interdependency result in specific rules of the game, discourses and the availability and division of resources. In Sect. 2.3.3, the process of structuration will be further explained as a process of morphogenesis and elaboration, which refers to those processes, which tend to elaborate or change a system's given form, structure or state (Buckley in Archer 2010a: 274; Archer 2010a, b, 2014). In the process of stabilization, already formed governance arrangements constrain the agency of the involved actors into adopting certain discourses, rules and resources, while actors defend the existing governance arrangements based on their interests, relations of interdependence, etc. In Sect. 2.3.3, this will be further elaborated as morphostasis (and maintenance), which refers to processes in a complex system that tend to preserve governance arrangements unchanged (Archer 2010a: 274).

Having said that institutionalization is about change and stability, dominant strands of new institutionalism, such as Rational Choice Institutionalism, Sociological Institutionalism and Historical Institutionalism emphasize the process of stability, for example the forming of institutions, while facing problems to explain institutional change. According to Mahoney and Thelen (2010) institutional change often occurs "when problems of rule interpretation and enforcement open up space for actors to implement existing rules in a new way" (Mahoney and Thelen 2010: 8). According to them ambiguity is a permanent feature of interpreting the rules and the way rules are instantiated in practice, even where rules are formalized. "Actors with divergent interests will contest the openings this ambiguity provides because matters of interpretation and implementation can have profound consequences for resource allocations and substantive outcomes" (Mahoney and Thelen 2010: 11). They distinguish four forms of institutional change: *displacement* (removal of existing rules and the introduction of new ones), *layering* (introduction of new rules on top of or alongside existing ones), *drift* (the changes impact of existing rules due to shifts in the environment) and *conversion* (the changed enactment of existing rules due to their strategic redeployment) (pp. 15–18). Drift and conversion are according to Hacker et al. (2015) neglected modes of change. The formal rules embodied in institutions remain constant but either the outcomes of these rules (drift) or the ways in which they are interpreted and used (conversion) change in politically consequential ways. In the case of drift, the trigger for change is *context discontinuity*, while in the case of conversion, the trigger is *actor discontinuity*. These forms of institutional change are one of the building blocks to develop the continuum of liquid institutionalization (see Sect. 2.6).

2.3.2 A Liquid Ontology

Liquid institutionalization is based on a liquid ontology. Ontology is the study of being and how we understand reality or the world around us (as external to humans (objectivism) or interpreted by humans (constructivism) (Toonen 2013). The reason why we need a liquid ontology to understand the reality of governance structures and processes at sea is related to the general societal and political processes of transformation, reflecting the changing relations between state, civil society and market on the one hand, and the characteristics of social and political activities at seas on the other (political modernization).

Firstly, contemporary societies are in transition. Social scientists labelled these processes of transformation, for example, as post-Fordism (Kumar 1995), post-modernism (Albrow 1996), late or reflexive modernity (Ulrich Beck et al. 1994), liquid modernity (Bauman 2000). Beck's risk society (Ulrich Beck 1992, 1994) and Castells network society (Manuel Castells 2010a, b, c) analyse in detail the new ordering principles of contemporary societies, such as risks and uncertainties, networks and flows in the stage of second modernity. In this second stage of modernity "the indissoluble link of society and nation state is fundamentally broken with the emergence of a logic of *flows* including (…) the flows of risks (…). In such a situation modernity is radicalized, subjecting itself to reflexive processes. Second or reflexive modernization disenchants and dissolves its own taken-for-granted foundations" (Urry 2004). The idea of second modernity not only allowed Beck to consider reflexivity as central to the reconstitution of society but also to pose scenarios of collapsing boundaries that required new skills involving flexibility and negotiability (Lee 2011). According to Lee (2011: 660) to argue for reflexivity as the basis for change implies a consideration for the recursive effects of flexible rationality. "If the processes of differentiation were treated as solidifying modernity over time, then the recursiveness of reflexivity and multiplicity would only serve to undermine temporal linearity by confronting, challenging and reversing preceding developments" (Lee 2011: 660–661). In this sense, reflexivity is about the melting of solids and presents a state of fluidity and uncertainty because it involves de-differentiation of social roles and social spaces. This is what Bauman calls "liquid modernity" (Bauman 2000). We are living in a liquid modern society, which means that the nature of our lives is fluid and always in flux (Ligocki 2019: 81). The fluidity of the world has enabled a complex breakdown of social networks, posture of insensitivity and a disposability of both people and things. "What distinguishes liquid modernity from early modernity is the lack of stable institutions. There is no condition; everything is process" (Abrahamson 2004: 171). With liquid modernity, Bauman wants to understand the alienation of progress and the unbearable human suffering and injustice (Lee 2005). In liquid modernity there is a state of ambient insecurity, anxiety and fear (Bauman and Haugaard 2008). "The disintegration of the social network, the falling apart of effective agencies of collective action is often noted with a great deal of anxiety and bewailed as the anticipated 'side effect' of the new lightness and fluidity of the increasingly mobile, slippery, shifty, evasive and fugitive power" (Bauman 2000: 14). Power has become truly exterritorial (Bauman 2000: 11). The difference with reflexive modernists is that Bauman treats liquidity as irreversible, while reflexive modernists consider the breakdown of the differentiated order as possibly leading to new forms of solidity. Lee (2011: 651) criticizes the irreversibility of liquidity because "liquidity itself can be considered an impermanent process and therefore its limits can be identified."

Secondly, at sea many things are moving and in flux, such as water (waves and currents), and resources (fish), but liquidity not only refers to the physical characteristics of seas, but especially to the liquidity of maritime activities (such as fishing and navigation/shipping). Liquidity of maritime activities at sea refers on the one hand to the mobility of activities and resources out of sight. This makes these activities invisible, which complicates their control and governing. This is what Anderson and Peters

(2014) and Steinberg and Peters (2015) call a wet or fluid ontology,[1] in which everything is in flux, changeable, processual and in a constant state of becoming, which they capture in the concepts hydrosphere, liquidity and dynamism. On the other hand, seas and oceans are social constructions. For example Steinberg (2001) views oceans "as a *social* space, a space *of* society" (Steinberg 2001: 6). His "territorial political economy perspective" emphasizes uses, regulations and representations of a space's social construction,[2] instead of a traditional perspective, which conceptualizes oceans as a space of resources, a transport surface and as a battleground or "force-field" (Steinberg 2001: 11–18). "The sea remains (…) a space constructed amidst competing interests and priorities, and it will continue to be transformed amidst social change" (Steinberg 2001: 207). By presenting different ideal-type constructions of oceans (as a social space between societies (Indian Ocean), as perceived and managed as an extension of land space (Micronesian), and as a non-possessible, but legitimate arena for expressing and contesting social power (Mediterranean)) Steinberg sketched the outlines of liquidity as expressed in the final sentences of his book. "Amidst the dynamic processes of global political economy, the sea is a space of contradictions and alternatives, of images and laws, of labour and dreams. The sea never stops moving" (Steinberg 2001: 210).

I define *liquid ontology* as the study of the nature of social being at sea, the kind of things that exist in the reality at sea, the conditions of existence, the relations of interdependency and the standards that have to be met for something to fully exist at sea (based on (Elder-Vass 2013: 15–16; Kauppi 2018: 28–29; Scott and Marshall 2009). Liquid ontology is what

---

[1] A fluid ontology is "promoting a knowledge of the world which is *neither* 'land' biased *nor* 'locked' to static and bounded interpretations of space, but rather one that conceives of our (water) world as one which is in flux, changeable, processual and in constant state of becoming" (Anderson and Peters 2014: 4–5).

[2] "The political-economic logic and structures of a given society lead social actors to implement a series of uses, regulations and representations in specific spaces, including ocean-space. Once implemented in a particular space, each aspect of the social construction (each use, regulation, and representation) impacts the others, effectively creating a new "nature" of that space. This "second nature" is constructed both materially and discursively, and it is maintained through regulatory institutions. Finally, the social construction of space impacts the material organization of society, both directly and indirectly through its re-construction of the nature that provides the foundation for social organization" (Steinberg 2001: 21–22).

Kauppi (2018: 36–41) calls a reflexive ontological framework.[3] It is important to distinguish two forms of liquidity. Firstly, related to materiality, resources and activities, these are tangible forms of liquidity such as waves, the water column, fish (as a resource), and shipping or tourism (as activities). Secondly, intangible forms of liquidity related to the "invisibility" of activities and the ambiguity of governing. Liquid institutionalization as developed in the theoretical framework is based on the intangible forms of liquidity.

2.3.3   Understanding Change and Stability in (Liquid) Institutionalization

Processes of institutionalization are about change and stability as expressed in the processes of structuration and stabilization. To understand these processes we build on theories in which "action" and "structure" presuppose one another and in which structural patterning is inextricably grounded in practical interaction, such as the structuration theory (Giddens 1984) and the morphogenetic approach (Archer 2010b), and on theories which explain change and transformation. Although Giddens accepts the bracketing of either structure and agency he nonetheless adopts an agency-centred analysis (Arts and van Tatenhove 2006). Liquid institutionalization is based on an analytical dualism of structure and agency, in which structure and agency are of a different order and logic. Examples of these dialectical approaches are Jessop's strategic-relation approach (Jessop 1990) and Archer's morphogenetic approach (Archer 1995, 2010a, b). Both theories make a clear distinction between structure and agency (analytical dualism) in which action takes place within a pre-existing structural context, but "people are not puppets of structures because they have their own emergent properties which mean they either reproduce or transform social structures rather than create them" (Archer 1995 in McAnulla 2002). According to Archer action "is ceaseless and essential both to the continuation and further elaboration of the system, but subsequent interaction will be different from earlier action because

---

[3] This framework analyses institutions as embodied entities involving individual and collective action, the political reality consists of interdependent actors and institutions, reality is not natural but there is always choice and agency, is a relations approach and the inclusive ontological position emphasizes "the ties between the macro and the micro, institutions and power and actions of individuals and groups in more or less structured social spheres" (Archer 2010c: 36).

conditioned by the structural consequences of that prior action" (Archer 2010b: 228). Additionally, the morphogenetic approach is sequential, because it "is dealing in endless cycles of—structural conditioning/social interaction/structural elaboration—thus unravelling the dialectical interplay between structure and action" (Archer 2010b: 228).[4] The morphogenetic argument that structure and action operate over different time periods is based on two propositions:

1. that structure logically predates the action(s) which transform it,
2. that structural elaboration logically postdates those actions (Archer 2010b: 238).

This makes it possible to analyse change and radical social transformation (Archer 2013). By focusing on social morphogenesis it is (1) possible to deal with "'those processes which tend to elaborate or change a system's given form, structure, or state' in preference to morphostatic processes, 'that tend to preserve or maintain a system's form, organization or state'" (Buckley 1967, as cited by Archer 2013: 2), and (2) in quest of a generative mechanism accounting for change and transformation.

Archer's morphogenetic approach consists of cycles of changes that can be summarized as follows: the social interactions of agents are affected by the structural and cultural conditions (conditioning), which are the result of past actions (T^1—marking the analytical starting point of processes of change). Because there is not a deterministic relation between structure and agency; actors have some degree of independent power to affect events in social interactions (T^2–T^3). This might result in changing conditions. This process of elaboration (or reproduction) is a result that no actor could foresee beforehand but emerges as the outcome of conflict or compromise. The changed structural and cultural context marks the beginning of a new circle of change (T^4 is the new T^1 of the next cycle of conditioning–interactions–elaboration). But in many cases, events leave conditions relatively unchanged, or actions fail to bring about desired changes (process of morphostasis: conditioning–interactions–maintenance) (Archer 2010b) (see Fig. 2.2)

---

[4] Hence Giddens's whole approach turns on overcoming the dichotomies which the morphogenetic perspective retains and utilizes—between voluntarism and determinism, between synchrony and diachrony, and between individual and society (Archer 2010b: 228).

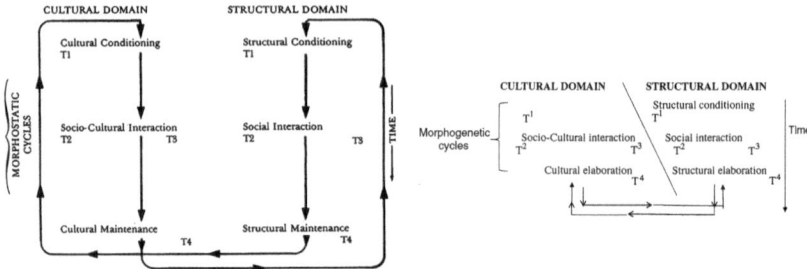

Fig. 2.2 Societal morphostasis (left) and societal morphogenesis (right) in (Archer 2013: 7)

2.4   Reflexivity in (Liquid) Institutionalization

Liquid institutionalization is about the continuous processes of structuration. Its ambiguity makes it possible to continuously negotiate the rules of the game. To understand this permanent state of negotiation and change in this section I conceptualize liquid institutionalization as a form of reflexivity.

In general, reflexivity is about self-confronting (Ulrich Beck 2006) and the ability of actors to turn or bend back on themselves (Hendriks and Grin 2007: 334). The transition to reflexive modernity brings with it the challenge of developing a new logic of action and decision, which no longer finds its orientation in the principle "either this or that" but rather in the principle "this and that both" (Beck 2006: 33). According to Archer, the distinguishing feature of reflexivity is "that it has the self-referential characteristic of 'bending-back' some thought upon the self, such that it takes the form of *subject-object-subject*" (Archer 2010c: 2). Archer defines reflexivity as "the regular exercise of the mental ability shared by all normal people to consider themselves in relation to their (social) contexts and vice versa" (Archer 2007: 4)

To understand the role of reflexivity in liquid institutionalization we start with Donati's conceptualization of reflexivity from relational sociology, who extends Archer's definition of reflexivity by including the distinction between personal and social reflexivity and systemic reflectivity (Donati 2010a). According to Donati, reflexivity is not only a reactive process (as described by Beck et al. 1994) nor self-confrontation, but "a capacity for reorientation and redirection, helping to build up new structures (...)" (Donati 2010a: 144). Reflexivity can be understood as a social relation between what Donati calls "Ego" and "Alter" within a social context (Donati 2010a: 145). Personal reflexivity (or internal conversation) is when Alter is Ego's self, while social reflexivity refers to an interactive situation where Alter is another person, and actors gain insights from these interactions (Donati 2010b, 2011). Next to personal and social reflexivity, Donati also defines systemic reflectivity,[5] referring to the sociocultural structures and their interactive parts, that is if reflexivity has pertained to structural and cultural conditions, and becomes an elaboration emerging from personal and social reflexivity in the social interactions (Ego and Alter are part of a system) (Donati 2010a).

In processes of institutionalization the behaviour of actors and interactions between these actors are mediated by different modes of reflexivity under the constraints of structural and cultural conditioning (T^1). During T^1, dominant governance arrangements are stabilized, in terms of dominant coalitions, rules, resources and discourses, due to processes of societal, economic and political transformations. These processes of political modernization, such as an emerging network state at sea and related changes in authority (van Tatenhove 2016), the rise of informational governance and digitalization (Castells 2004; Toonen 2013), and the rise of market-driven governance and authority (Miller and Bush 2015; Bailey et al. 2016) shape the structural and cultural conditions for blue governance arrangements. Interactions in blue governance arrangements during T^2–T^3 do not take place in a social and political vacuum but within the structural and cultural conditions in which these blue governance arrangements developed and stabilized. This is reflected in the structural properties—rules, resources, discourses—of these (dominant) blue governance arrangements (rules, resources and discourses). The structural conditions structure also possible forms of reflexivity and the capacity of actors to do otherwise. The reflexivity that is activated during interactions (T^2–T^3) is an expression of *agency*, which mediated the structures within which they operate (Donati 2010a: 155). The passage from T^2 to T^3 depends on the enabling and constraining conditions of the structural properties of

---

[5] Donati speaks of system reflectivity instead of system reflexivity. "*Reflection* is a self-referential operation of an individual mind which bends back on itself within itself" (Donati 2010a: 146). *Reflexivity* "is a relational operation on the part of an individual mind in relation to an 'Other' who can be internal (the Ego as an Other) or external (Alter) but who also takes the social context in account" (Donati 2010a: 147).

dominant blue governance arrangement and the transformative capacity of emerging governance arrangements to challenge dominant ones. The new conditions emerging at T^4 depend on the extent to which reflexivity has been developed between T^2 and T^3 (Toonen and van Tatenhove 2020). While Donati (2010b) makes a distinction between personal and social reflexivity, there is no clear insight in the extent to which the conditions of reflexivity are affected.

To understand if and how processes of institutionalization are becoming reflexive and to which extent one can speak of structural elaboration or maintenance, I will make use of previously developed modes of reflexivity (van Tatenhove 2017), that is structural and performative reflectiveness and reflexivity. *Structural reflectiveness* refers to the ability of actors to use rules and resources from different institutional settings within a given discursive space of a policy domain, but actors are not able to change the rules of the game. The dominant form of mobilization of actors is action-oriented within an existing governance setting. The conditions remain relatively unchanged (morphostasis). *Performative reflectiveness* refers to the ability of actors to challenge the discursive space of a governance arrangement (performative mobilization) (Pestman 2001). This could result in for example alternative discourses, and related new coalitions, rules and resources existing side by side with the existing governance arrangement, but existing institutional rules and power relations (polity) are not challenged. *Reflexivity* refers to the situation when actors both challenge the existing discursive space of a policy domain, and are able to change the institutional rules (structural congruence) (Boonstra 2004), which thus refers to a process of morphogenesis (structural and cultural elaboration).

Blue governance arrangements are reflexive if coalitions of actors are capable to confront, question, and amend existing rules, and have the ability to change the rules of the game and to mobilize resources (given their dispositional power position) based on alternative discourses. Reflexivity in processes of institutionalization refers to the capacity of actors to govern and to induce institutional change (i.e. to change the processes of structuration and stabilization) by challenging the existing discursive spaces of marine governance arrangements (performative mobilization), and to activate and to use rules and resources from different rule systems and layers of government (van Tatenhove 2017). In this sense, the dynamic process of institutionalization is driven by agency (reflexivity) and structural conditioning (networked polity and related power structures). Liquid institutionalization at sea is not planned and designed, but is what Beck and Grande (2007: 6) call "institutionalized improvisation."

2.5   POWER IN LIQUID INSTITUTIONALIZATION

Power is the "organisational and discursive capacity of agencies, either in competition with one another or jointly, to achieve outcomes in social practices, a capacity which is however co-determined by the structural power of those social institutions in which these agencies are embedded" (Arts and van Tatenhove 2005: 347). Power is a multi-layered phenomenon, consisting of relational, dispositional and relational power.

Relational power refers to the capability of actors to achieve outcomes through interactions. Such power is not an amulet held by a single actor but is always constituted in and exerted through social relations (Elias 1970; Goehler 2000; van Tatenhove et al. 2010). Relational power posits actors, resources, outcomes and interactions as the constitutive elements of power at this level. However, the capacities of actors to influence outcomes are unevenly distributed. This may be due to differences in their psychological or verbal capacity, to an unequal distribution of resources or to an unequal access to resources (van Dijk et al. 2011). In other words relational power is constituted and exerted in social relations and interactions in which interdependent actors are capable of achieving outcomes, by mobilizing and deploying the available resources (Arts and van Tatenhove 2005). Actors have influence when they are able to change policy outcomes.

Dispositional power shapes an "agency's capacity to act" (Clegg 1989: 84). Organizational rules and an unequal distribution of resources define and position actors vis-à-vis each other. These positions codetermine what agents may achieve in terms of relational power. Rules and resources mediate the process of positioning (Arts and van Tatenhove 2005). The unequal distribution of resources, how they are applied, the positioning of actors in planning processes in the face of external constraints (Coleman 1977) are all factors that are the result of structural power.

Structural power refers to the way in which macro-societal structures shape the nature and conduct of agents. This is what (Giddens 1984) calls orders of signification, legitimization and domination that "materialize" in discourses as well as in political, legal and economic institutions in societies. "Mediated by these discourses and institutions, (collective) agents give meaning to the social world, consider some acts and thoughts legitimate, and others not, and are enabled or constrained to mobilise resources to achieve certain outcomes in social relationships" (Arts and van Tatenhove 2005: 351). Because of the "structured asymmetries of resources," given by any order of domination, actors have uneven access to the use of resources.

By defining processes of institutionalization as multi-level power games, power should be understood and analysed at the three interconnected levels and their interplay. At the level of policy practices (relational power); at the level of the blue governance arrangements (dispositional power) and at the level of political modernization (structural power). This multi-level power framework makes clear that, power is not restricted to the mobilization of resources or the achievement of outcomes by actors in interactions alone, but it also includes dispositional and structural phenomena. The analysis of power can start at each of the levels but should take in consideration the other two levels.

To understand the interplay of these levels of power processes and the role of power holders in processes of liquid institutionalization at sea, I will combine the multi-layered framework with Castell's concept of network-making power and the holders of power (Castells 2009). Castells distinguishes four forms of power in the global network society; networking power; network power; networked power and network-making power. *Networking power* refers to the power of actors (included in the network) over actors outside the network and/or the power to exclude actors (Castells 2009; 42–43). *Network power* refers to standards or protocols which determine the rules to be accepted once in the network. Power "is exercised not by exclusion from the networks, but by the imposition of the rules of inclusion" (Castells 2009: 43). *Networked power* is about the operation of power in dominant networks. In general, this refers to the relational capacity of an actor to impose its will over another actor's will on the basis of the structural capacity of domination embedded in the institutions of society (Castells 2009: 44). According to Castells, traditional conceptions of power do not make sense in the network society, because "new forms of domination and determination are critical in shaping people's lives regardless of their will" (Castells 2009: 45). For Castells, the most crucial form of power is *network-making power*, which is based on two mechanisms: "(1) the ability to constitute networks(s), and to program/reprogram the network(s) in terms of the goals assigned to the network; and (2) the ability to connect and to ensure the cooperation of different networks by sharing common goals and combining resources, while fending off competition from other networks by setting up strategic cooperation" (Castells 2009: 45). Holders of the first power position are programmers, while holders of the second positions are switchers. In other words, network-making power is the power of programmers, based on their interests, norms and values, to program specific networks, while switchers have the control of connecting between different strategic networks, following the strategic alliances between the dominant actors of various networks (Castells 2011). Programming and switching are the power mechanisms in network-making power.

Combining the multi-layered power framework of relational, dispositional and structural power with the mechanisms and power holder positions in network-making power makes it possible to understand power processes and power positions in different forms of liquid institutionalization. Institutionalization is about continuous processes of change and stability, conceptualized in this chapter as structuration or elaboration (morphogenesis) and stabilization or maintenance (morphostasis). In situations in which the status quo successfully is defended against change institutionalization takes the characteristic of stabilization/maintenance. Powerful actors/coalitions are able to counteract change by defending successfully dominant discourses, rules and the division of resources. In processes of liquid institutionalization, understood as a continuous process of structuration/elaboration, the positioning of actors in developing governance arrangements is in flux under conditions of high ambiguity. In interactions actors make use of rules, change rules and mobilize (unequally distributed) resources due to different institutional orders (structural power). In liquid institutionalization relational and dispositional power are in a permanent state of flux and liquidity; actors have the possibility to change the structural properties (rules and resources) of different governance arrangements. In situations of high ambiguity actors do not have fixed positions, while the positioning vis-à-vis other actors is dependent on the network-making capabilities of actors in governance arrangements, that is which actors are able to "program" the governance arrangements in terms of discourses, coalitions and rules and resources (programmers) and which actors have the ability to bring coalitions and discourses of different governance arrangements together (switchers). Who the programmers are and who the switchers are in blue governance arrangements is an empirical question, and has to be answered for maritime activities within their specific institutional context. This will be analysed in the next chapters for Arctic shipping, deep seabed mining and transboundary regionalization.

2.6   A CONTINUUM OF TYPES OF LIQUID INSTITUTIONALIZATION

Based on the building blocks developed in the previous sections, this section defines liquid institutionalization in more detail, resulting in a continuum of types of liquid institutionalization. In general, as discussed previously, institutionalization is the formation and stabilization of governance arrangements in processes of structuration and stabilization. Governance arrangements develop within cultural and structural (institutional) contexts, which enable or constrain these processes of institutionalization, because the institutional context (formed in processes of political modernization) results in an unequal division of resources, dominant discourses, legitimate rules of the games and the formation of possible coalitions. Interactions between actors take place in and are enabled and/or constrained by the given institutional setting, which forms the relations of interdependence and the specific division of resources, legitimate rules and possible discourses. The outcomes of interactions are processes of structural structuration/elaboration or structural stabilization/maintenance. In the case of structural elaboration, the structural properties of (institutionalized) governance arrangements are challenged which could result in changing cultural and structural conditions (political modernization). The structural properties of governance arrangements are (re)produced in interaction, however changes in processes of political modernization are only indirectly related to agency, change in these structural conditions is "the *unintended* outcome of social interactions and outcomes in many governance arrangements across time and space" (Arts and van Tatenhove 2006: 38). In the case of structural maintenance, structural properties of governance arrangements could be challenged, but the structural and cultural conditions are preserved, and the given institutional order is maintained or stabilized.

Earlier I defined liquid institutionalization as an ongoing process of structuration in which blue governance arrangements are continuously negotiated, while in a permanent state of "flux and becoming" without a solid stabilization in institutionalized blue governance arrangements. This permanent state of "flux and becoming" is the consequence of governance contexts of high ambiguity and reflexivity. However, political processes and policy-making are not in processes of permanent flux. Outcomes of negotiations always present a moment of stabilization. Liquid institutionalization is therefore an analytical ideal type. An ideal type is "the construction of certain elements of reality in a logically precise conception" (Gerth and Wright Mills 1982: 59).This chapter will present a continuum of types of liquid institutionalization: Type I, Type II, Type III and Type IV (inspired by Hooghe and Marks 2003). These types express different forms of liquidity; frozen, gelatinous, syrupy and liquid, and are developed by combining the "variables" structuration/elaboration—stabilization/maintenance, the four forms of institutional change by Mahoney and Thelen (2010) and van Tatenhove's forms of reflectiveness and reflexivity.

Liquid institutionalization refers to a continuous process of institutional change. However, as discussed in Sect. 2.3.1 different forms of institutional change can be distinguished, such as layering, displacement, drift and conversion. In case of layering, new rules are attached to existing ones, thereby changing the ways in which the original rules structure behaviour. In case of displacement, new rules replace existing ones. This can be a sudden breakdown or dismantling of institutions due to shock-events (revolutions, natural disasters, etc.) or the result of more slow-moving processes when new institutional rules of institutions compete with existing ones, leading to a gradual displacement. Forms of more gradual institutional change are drift and conversion (see Sect. 2.3.1). In both forms, institutional rules remain formally the same, but in case of drift their impact changes because of external conditions, while conversion refers to a situation in which rules are interpreted and enacted in new ways.

Combining the building blocks developed in the Sects. 2.3.1, 2.3.3, and 2.4 results in a continuum of liquid institutionalization (see Fig. 2.3).

Type I (Frozen) forms the one end of the continuum, expressing no or little liquidity. Characteristic for this type of liquid institutionalization is stabilization/maintenance. Interactions (or events) do not change the (structural) conditions nor fail to bring about desired changes. This process of stabilization/structural maintenance reflects a situation in which existing (blue) governance arrangements are defended against change; in interactions where actors are not able to change the rules, resulting in structural maintenance (instead of elaboration). Structural maintenance reflects no institutional change; existing rules are defended against change, and if new rules are developed, they do not displace nor convert existing ones but are layered upon existing ones. This type of institutionalization

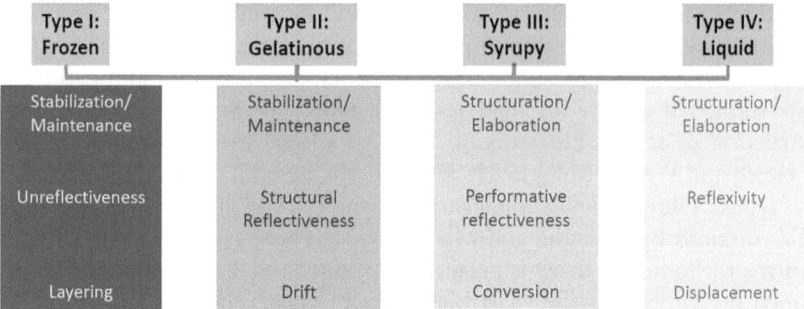

Fig. 2.3   Continuum of liquid institutionalization

and institutional change is related to situations of unreflectiveness (neither the discursive spaces of a policy domain nor the institutional rules are challenged). In Type I the structural conditions or the orders of signification, domination, and legitimization (structural power) determine the unequal division and availability of resources and the positioning of actors in blue governance arrangements, in this way structuring the possible interactions, which can take place given structural power.

The other end of the continuum, *Type IV (Liquid)* expresses total liquidity. Actors (inter)act within specific structural circumstances; in these interactions they alter these circumstances by changing the rules of the game. This process of structural elaboration is driven by reflexivity in which actors can both challenge the existing discursive space of a policy domain and change the institutional rules.

Type II (Gelatinous) and *Type III (Syrupy)* are in-between types of liquid institutionalization. Although *Type II (Gelatinous)* represents maintenance, institutional changes and flux are the result of structural reflectiveness and drift. In interactions actors have the ability to strategically use rules from different institutional settings/layers within the given discursive space of a policy domain (structural reflectiveness). The governance arrangement is embedded in its institutional context and actors are able to mobilize rules and resources from different layers, but they are not able to change the rules of the game (van Tatenhove 2017: 790). In this type of liquid institutionalization, change and flux are the result of structural reflectiveness combined with drift. Drift is the unintended consequence of not updating existing institutions, or the intended strategy of actors to block adaptation of institutions to changing circumstances (Hacker et al.

2015; van der Heijden and Kuhlmann 2017). In the words of Hacker et al. (2015: 184) drift is "the failure of relevant decisions makers to update formal rules when shifting circumstances change the social effects of those rules in ways that are recognized by at least some political actors." An example of institutional drift is the impeded capacity of the Great Barrier Reef National Marine Park Authority to effectively manage issues arising from the increase industrialization and urbanization of coastal Queensland (Kelly et al. 2019).

Type III (Syrupy) of liquid institutionalization is a form of institutional liquidity as structural elaboration based upon performative reflectiveness and conversion. Institutional change as conversion refers to situations in which rules formally remain the same, but are integrated and enacted in new ways; existing institutions are redirected to new purposes driving change in the role they perform and/or the function they sense (Thelen 2003). Besides conversion, change in Type III is also the result of performative reflectiveness. In the case of performative reflectiveness, actors challenge the discursive space of a governance arrangement. This could lead to an alternative discourse challenging the dominant discourse, resulting in new coalitions, rules and resources, and thus the institutionalization of an alternative governance arrangement side by side with the existing governance arrangement (van Tatenhove 2017: 790). An example of Type III liquid institutionalization, in which change is the result of conversion, performative reflectiveness and structural elaboration is the possible conversion of Advisory Councils in EU fisheries management into Integrated Marine Governance Councils (van Tatenhove 2011), due to a discursive change emphasizing the development of multiple maritime sectors at the regional sea level. Although actors do challenge the discursive space, the formulation of policy objectives, the definition of problems, and the solutions take place within the given institutional EU maritime rules.

The next chapters will analyse the institutionalization of Arctic Shipping, Deep Seabed Mining (DSM) and Transboundary Regionalization. The main question for each chapter is what kind of blue governance arrangements do institutionalize and what type of liquid institutionalization do they represent?

REFERENCES

Abrahamson, Peter. 2004. Review Essay Liquid Modernity. *Acta Sociologica* 47 (2): 171–179. https://doi.org/10.1177/0001699304043854.

Albrow, Martin. 1996. *The Gobal Age. State and Society Beyond Modernity*. Cambridge, UK: Polity Press.

Anderson, Jon, and Kimberley Peters. 2014. *Water Worlds: Human Geographies of the Ocean*. Ashgate Publishing Ltd.

Archer, Margaret S. 1995. *Realist Social Theory: The Morphogenetic Approach*. Cambridge, UK: Cambridge University Press.

———. 2007. *Making Our Way through the World: Human Reflexivity and Social Mobility*. Cambridge: Cambridge University Press.

———. 2010a. Routine, Reflexivity, and Realism. *Sociological Theory* 28 (3): 272–303. https://doi.org/10.1111/j.1467-9558.2010.01375.x.

———. 2010b. Morphogenesis Versus Structuration: On Combining Structure and Action. *British Journal of Sociology* 61 (Suppl. 1): 225–252. https://doi. org/10.1111/j.1468-4446.2009.01245.x.

Archer, Margaret Scotford. 2010c. *Conversations About Reflexivity*. Routledge.

Archer, M.S. 2013. *Social Morphogenesis*. Springer.

———., ed. 2014. *Late Modernity. Trajectories towards Morphogenic Society*. New York/Dordrecht/London: Springer. https://doi. org/10.1007/978-3-319-03266-5.

Arts, Bas, and Jan P.M. van Tatenhove. 2005. Policy and Power: A Conceptual Framework between the 'Old' and 'New' Policy Idioms. *Policy Sciences* 37 (3–4): 339–356. https://doi.org/10.1007/s11077-005-0156-9.

———. 2006. Political Modernisation. In *Institutional Dynamics in Environmental Governance*, 21–43. Dordrecht: Springer Netherlands. https://doi.org/1 0.1007/1-4020-5079-8_2.

Arts, Bas, Pieter Leroy, and Jan P.M. van Tatenhove. 2006. Political Modernisation and Policy Arrangements: A Framework for Understanding Environmental Policy Change. *Public Organization Review* 6 (2): 93–106. https://doi. org/10.1007/s11115-006-0001-4.

Bailey, Megan, Simon R. Bush, Alex Miller, and Momo Kochen. 2016. The Role of Traceability in Transforming Seafood Governance in the Global South. *Current Opinion in Environmental Sustainability*. https://doi.org/10.1016/j. cosust.2015.06.004.

Bauman, Zygmunt. 2000. *Liquid Modernity*. Cambridge, UK: Polity Press.

Bauman, Zygmunt, and Mark Haugaard. 2008. Liquid Modernity and Power: A Dialogue with Zygmunt Bauman. *Journal of Power* 1 (2): 111–130. https:// doi.org/10.1080/17540290802227536.

Beck, Ulrich. 1992. *Risk Society: Towards a New Modernity*. Ed. Mark Ritter. *Nation*. Vol. 2. Theory, Culture & Society Series. Sage. https://doi. org/10.2307/2579937.

———. 1994. The Reinvention of Politics: Towards a Theory of Reflexive Modernization. In *Reflexive Modernization*, ed. U. Beck, A. Giddens, and S. Lash, 1–55. Cambridge: Polity Press.

————. 2006. Reflexive Governance: Politics in the Global Risk Society. In *Reflexive Governance for Sustainable Developement*, ed. Jan-Peter Voß, Dierk Bauknecht, and Rene Kemp, 31–56. Celtenham, UK: Edward Elgar Publishing Limited.

Beck, Ulrich, and Edgar Grande. 2007. *Cosmopolitan Europe*. Cambridge, UK; Malden, USA: Polity Press.

Beck, Ulrich, Anthony Giddens, and Scott Lash. 1994. *Reflexive Modernization: Politics, Tradition and Aesthetics in the Modern Social Order*. Stanford University Press.

Boonstra, F.G. 2004. Laveren Tussen Regio's En Regels. Verankering van Beleidsarangementen Rond Plattelands Ontwikkelingen in Noordwest-Friesland, de Graafschap En Zuidwest Salland. http://hdl.handle.net/2066/67439.

Castells, Manuel. 2004. Informationalism, Networks, and the Network Society: A Theoretical Blueprint. In *The Network Society*, 3–45. Cheltenham: Edward Elgar Publishing. https://doi.org/10.4337/9781845421663.00010.

Castells, M. 2009. *Communication Power*. Oxford: Oxford University Press.

————. 2010a. *The Rise of the Network Society: With a New Preface, Volume I: Second Edition*. 1st ed. Wiley. https://doi.org/10.1002/9781444319514.

————. 2010b. *The Power of Identity*. Wiley-Blackwell.

————. 2010c. *End of Millennium: With a New Preface, Volume III, Second Edition*. 1st ed. Chichester: Wiley. https://doi.org/10.1002/9781444323436.

————. 2011. A Network Theory of Power. *International Journal of Communication* 5 (1): 773–787.

Clegg, Stewart R. 1989. *Frameworks of Power*. London: SAGE Publications Ltd.

Coleman, J. 1977. Notes on the Study of Power. In *Power, Paradigms and Community Research*, ed. R.J. Liebert and A.W. Imerskein, 183–198. London: Sage.

Donati, Pierpaolo. 2010a. Reflexivity After Modernity. From the Viewpoint of Relational Sociology. In *Conversations About Reflexivity*, ed. Margaret S. Archer, 144–164. London and New York: Routledge.

————. 2010b. *Relational Sociology*. 1st ed. Hoboken: Routledge. https://doi.org/10.4324/9780203860281.

————. 2011. Modernization and Relational Reflexivity. *International Review of Sociology* 21 (1): 21–39. https://doi.org/10.1080/03906701.2011.544178.

Elder-Vass, Dave. 2013. *The Reality of Social Construction*. Cambridge, UK: Cambridge University Press.

Elias, N. 1970. *Was Ist Soziologie?* Munich: Juventa Verlag.

Gerth, H.H., and C. Wright Mills, eds. 1982. *From Max Weber: Essays in Sociology*. London, Boston, Melbourne and Henley: Routledge & Kegan Paul. Reprint.

Giddens, Anthony. 1984. *The Constitution of Society. Outline of the Theory of Structuration*. Oxford, UK: Polity Press.

Goehler, G. 2000. Constitution and the Use of Power. In *Power in Contemporary Politics*, ed. H. Goverde, P. Cerny, M. Haugaard, and H. Lentner, 41–58. London: Sage.

Hacker, Jacob S., Paul Pierson, and Kathleen Thelen. 2015. Drift and Conversion: Hidden Faces of Institutional Change. In *Advances in Comparative-Historical Analysis*, ed. James Mahoney and Kathleen Thelen, 180–208. Cambridge: Cambridge University Press. https://doi.org/10.1017/CBO9781316273104.008.

Hajer, Maarten. 1995. *The Politics of Environmental Discourse: Ecological Modernization and the Policy Process*. Oxford: Claredon Press.

Hendriks, Carolyn M., and John Grin. 2007. Contextualizing Reflexive Governance: The Politics of Dutch Transitions to Sustainability. *Journal of Environmental Policy and Planning*. https://doi.org/10.1080/15239080701622790.

Hooghe, Liesbet, and Gary Marks. 2003. Unravelling the Central State, But How? Types of Multi-Level Governance. *Reihe Politikwissenschaft* 97 (2): 233–243. https://doi.org/10.2307/3118206.

Jessop, Bob. 1990. *State Theory: Putting the Capitalist State in Its Place*. Cambridge, UK: Polity Press.

Kauppi, Niilo. 2018. *Toward a Reflexive Political Sociology of the European Union. Fields, Intellectuals and Politicians*. Palgrave Macmillan UK. https://doi.org/10.1007/978-3-319-71002-0.

Kelly, Christina, Geraint Ellis, and Wesley Flannery. 2019. Unravelling Persistent Problems to Transformative Marine Governance. *Frontiers in Marine Science* 6 (APR). https://doi.org/10.3389/fmars.2019.00213.

Kumar, K. 1995. *From Post-Industrial to Post-Modern Society. New Theories of the Contemporary World*. Oxford, UK: Blackwell Publishers Ltd.

Lee, Raymond L.M. 2005. Bauman, Liquid Modernity and Dilemmas of Development. *Thesis Eleven* 83 (1): 61–77. https://doi.org/10.1177/0725513605057137.

———. 2011. Modernity, Solidity and Agency: Liquidity Reconsidered. *Sociology* 45 (4): 650–664. https://doi.org/10.1177/0038038511406582.

Liefferink, Duncan. 2006. The Dynamics of Policy Arrangements: Turning Round the Tetrahedron. In *Institutional Dynamics in Environmental Governance*, ed. B. Arts and Pieter Leroy, 45–68. Springer. https://doi.org/10.1007/1-4020-5079-8_3.

Ligocki, Danielle T. 2019. Viewing Research for Social Justice and Equity Through the Lens of Zygmunt Bauman's Theory of Liquid Modernity. In *Research Methods for Social Justice and Equity in Education*, 81–89. Cham: Springer International Publishing. https://doi.org/10.1007/978-3-030-05900-2_7.

Mahoney, James, and Kathleen Thelen. 2010. A Theory of Gradual Institutional Change. *Explaining Institutional Change: Ambiguity, Agency, and Power*: 1–37. https://doi.org/10.1017/9780521134323.

McAnulla, S. 2002. Structure and Agency. In *Theory and Methods in Political Science*, ed. D. Marsh and G. Stoker 2nd, 271–291. New York: Palgrave Macmillan.

Miller, Alice M.M., and Simon R. Bush. 2015. Authority without Credibility? Competition and Conflict between Ecolabels in Tuna Fisheries. *Journal of Cleaner Production* 107: 137–145. https://doi.org/10.1016/j.jclepro.2014.02.047.

Pestman, Paul. 2001. *In Het Spoor van de Betuweroute. Mobilisatie, Besluitvorming En Institutionalisering Rond Een Groot Infrastructureel Project*. Amsterdam: Rozenberg Publishers.

Scott, John, and Gordon Marshall. 2009. *A Dictionary of Sociology*. 3rd ed. Oxford, UK: Oxford University Press. https://doi.org/10.1093/acref/9780199533008.001.0001.

Steinberg, Philip E. 2001. *The Social Construction of the Ocean*. Cambridge, UK: Cambridge University Press.

Steinberg, Philip, and Kimberley Peters. 2015. Wet Ontologies, Fluid Spaces: Giving Depth to Volume through Oceanic Thinking. *Environment and Planning D: Society and Space* 33 (2). https://doi.org/10.1068/d14148p.

Thelen, Kathleen. 2003. How Institutions Evolve. Insights from Comparative Historical Analysis. In *Comparative Historical Analysis in the Social Sciences*, ed. James Mahoney and Dietrich Rueschemeyer, 208–240. Cambridge, UK: Cambridge University Press.

Toonen, Hilde M. 2013. *SEA@SHORE—Informational Governance in Marine Spatial Conflicts at the North Sea*. Wageningen University.

Toonen, Hilde M., and Jan P.M. van Tatenhove. 2020. Uncharted Territories in Tropical Seas? Marine Scaping and the Interplay of Reflexivity and Information. *Maritime Studies* 19 (3): 359–374. https://doi.org/10.1007/s40152-020-00177-z.

Urry, John. 2004. Introduction: Thinking Soeciety Anew. In *Conversations with Ulrich Beck*, ed. Ulrich Beck and Johannes Willms, 1–10. Oxford: Polity Press.

van der Heijden, Jeroen, and Johanna Kuhlmann. 2017. Studying Incremental Institutional Change: A Systematic and Critical Meta-Review of the Literature from 2005 to 2015. *Policy Studies Journal* 45 (3): 535–554. https://doi.org/10.1111/psj.12191.

van Dijk, Judith, Maarten van der Vlist, and Jan P.M. van Tatenhove. 2011. Water Assessment as Controlled Informality. *Environmental Impact Assessment Review* 31 (2): 112–119. https://doi.org/10.1016/j.eiar.2010.04.009.

van Leeuwen, Judith, and Jan P.M. van Tatenhove. 2010. The Triangle of Marine Governance in the Environmental Governance of Dutch Offshore Platforms. *Marine Policy* 34 (3): 590–597. https://doi.org/10.1016/j.marpol.2009.11.006.

van Leeuwen, Judith, Luc van Hoof, and Jan P.M. van Tatenhove. 2012. Institutional Ambiguity in Implementing the European Union Marine Strategy Framework Directive. *Marine Policy* 36 (3): 636–643. https://doi. org/10.1016/j.marpol.2011.10.007.

van Tatenhove, Jan P.M. 1993. *Milieubeleid Onder Dak? Beleidsvoeringsprocessen in Het Nederlandse Milieubeleid in de Periode 1970–1990, Nader Uitgewerkt Voor de Gelderse Vallei.* Wageningen: Pudoc.

———. 1998. Political Modernization and Environmental Policy. In *L'environnement Au XXIe Siecle. The Environment in the 21st Century. Volume 1 LesEnjeux. The Issues*, ed. Jacques Theys, 475–487. Paris: GERMES.

———. 1999. Political Modernisation and the Institutionalisation of Environmental Policy. In *European Discourses on Environmental Policy*, ed. M. Wissenburg, G. Orhan, and U. Collier, 59–78. Aldershot: Ashgate.

———. 2011. Integrated Marine Governance: Questions of Legitimacy. *MAST* 10 (1): 87–113.

———. 2013. How to Turn the Tide: Developing Legitimate Marine Governance Arrangements at the Level of the Regional Seas. *Ocean and Coastal Management* 71: 296–304. https://doi.org/10.1016/j.ocecoaman.2012.11.004.

———. 2016. The Environmental State at Sea. *Environmental Politics* 25 (1): 160–179. https://doi.org/10.1080/09644016.2015.1074386.

———. 2017. Transboundary Marine Spatial Planning: A Reflexive Marine Governance Experiment? *Journal of Environmental Policy and Planning* 19 (6): 783–794. https://doi.org/10.1080/1523908X.2017.1292120.

van Tatenhove, Jan P.M., and Pieter Leroy. 2000. The Institutionalisation of Environmental Politics. In *Political Modernisation and the Environment. The Renewal of Environmental Policy Arrangements*, ed. Jan P.M. van Tatenhove, Bas Arts, and Pieter Leroy, 17–33. Dordracht/Boston/London: Kluwer Academic Publishers.

van Tatenhove, Jan P.M., Bas Arts, and Pieter Leroy. 2000. In *Political Modernisation and the Environment. The Renewal of Environmental Policy Arrangements*, ed. Jan P.M. van Tatenhove, Bas Arts, and Pieter Leroy. Dordracht/Boston/London: Kluwer Academic Publishers.

van Tatenhove, Jan P.M., Jurian Edelenbos, and Pieter Jan Klok. 2010. Power and Interactive Policy-Making: A Comparative Study of Power and Influence in 8 Interactive Projects in the Netherlands. *Public Administration* 88 (3): 609–626. https://doi.org/10.1111/j.1467-9299.2010.01829.x.

van Tatenhove, Jan P.M., Paulina Ramírez-Monsalve, Eira Carballo-Cárdenas, Nadia Papadopoulou, Chris J. Smith, Lieke Alferink, Kristen Ounanian, and Ronan Long. 2021. The Governance of Marine Restoration: Insights from Three Cases in Two European Seas. *Restoration Ecology* 29 (S2). https://doi. org/10.1111/rec.13288.

CHAPTER 3

Arctic Shipping

Abstract This chapter presents and analyses the institutionalization of Arctic shipping governance arrangements. It is expected that in the nearby future three Arctic shipping governance arrangements will emerge, related to the opening of three possible shipping routes: the Northeast Passage/Northern Sea Route, the Northwest Passage and the Transpolar Sea Route. These three Arctic shipping governance arrangements institutionalize within the dominant discursive space that shipping in the Arctic is a legitimate activity under the conditions and rules of environmental protection and human safety as formulated in the Polar Code. These governance arrangements differ in types of shipping, the actors involved and the availability of resources, related to the geographical area of the route, and the natural circumstances. The chapter concludes that the future liquid institutionalization of Arctic shipping governance arrangements shows characteristics of Type II liquid institutionalization.

Keywords Arctic shipping; Northeast Passage; Northern Sea Route; Northwest Passage; Transpolar Sea Route; Arctic governance setting; Polar Code; Arctic Council; Arctic states; China

3.1   INTRODUCTION

On 23 March 2021, the 600-metre-long containership "Ever Given" (part of the Evergreen fleet, operated by the Taiwanese company Evergreen Marine) was stranded in the Suez Canal due to a desert storm. The Dutch dredging company and offshore contractor Royal Boskalis Westminster N.V. managed to pull the ship loose on Monday, 29 March 2021. The ship blocked the Suez Canal for almost a week causing a serious blockage in a key artery of the world economy: 12% of the world trade is dependent on the Suez Canal route. On 6 July 2021, the ship weighed anchor after Egypt signed a compensation deal with its owners and insurers.[1]

In 2017, China presented three marine economic passages connecting Asia with Africa, Oceania, Europe and beyond, in a bid to advance maritime cooperation under the Belt and Road Initiative.[2] China wants to designate three "blue economic passages,"[3] one of them is the Polar Silk Route (PSR) via the Arctic Ocean. Although this route is not yet well explored, China's Arctic Policy states that the PSR "facilitates connectivity and sustainable economic and social development of the Arctic" (Tianming et al. 2021) by opening up an economic passage between China and Europe through the seas north of Russia. According to Tianming et al. (2021) before "the COVID-19 outbreak hit the world; China had planned to redirect up to 1% of its maritime trade to the PSR by the early 2020s."

These examples not only show the increasing importance of maritime transport for the global economy, but also the ongoing dependence on nineteenth-century connecting canals, such as the Suez Canal, the Panama Canal, and the Strait of Malacca, which act as the arteries of global trade. Shipping is a global maritime activity and ship owners, insurance

---

[1] The Suez Canal Authority (SCA) "has been seeking compensation from the Ever Given's Japanese owner Shoei Kisen for the cost of the salvage operation, damage to the canal's banks and other losses. The SCA initially asked for $916m compensation, including $300m for a salvage bonus and $300m for loss of reputation. But UK Club which insured Shoei Kisen for third-party liabilities—rejected the claim, describing it as 'extraordinarily large' and 'largely unsupported'. The SCA later lowered its demand to $550m. The final settlement, which has not been revealed, was agreed a few days ago and signed on Wednesday (07/07/2021, JvT) to coincide with the ship's release" (https://www.bbc.com/news/world-middle-east-57746424, visited 8 July 2021).

[2] http://english.www.gov.cn/state_council/ministries/2017/06/21/content_281475692760102.htm (visited 1 April 2021).

[3] The China-Indian Ocean-Africa-Mediterranean Sea Blue Economic Passage; the China-Oceania-South Pacific Blue Economic Passage; and a passage to Europe via the Arctic Ocean.

companies and states are looking for the most economic profitable and safe shipping routes to transport resources and goods all over the world. In these globalized shipping networks, countries like China have a growing influence and power position in defining the conditions of shipping routes. The increase of global maritime transport across the world's oceans and seas not only has economic implications, but also environmental and spatial consequences. Maritime transport contributed 3% to the global emissions of Greenhouse gasses (IMO 2014). The emission of GHG results in warming-up of the earth. An unintended consequence of climate change, more specifically the melting of sea ice in the Arctic Ocean, provides new opportunities for countries and the maritime transport sector to develop new shipping routes and maritime activities (such as port development and tourist hubs) in the Arctic region.

This chapter presents an analysis of the institutionalization of Arctic shipping (blue) governance arrangements. With the opening of shipping routes, due to a diminishing of the extent and volume of Arctic sea ice (Keil 2018), the governing of Arctic shipping has become both timely, and politically, socially and scientifically relevant. Arctic shipping is regulated by a patchwork of governance structures and regulations, with navigation taking place both in the territorial waters of Arctic states (Russia, Canada, the USA, Norway, Finland, and Denmark) and in the high seas (the Areas Beyond National Jurisdiction—ABNJ). Within their Exclusive Economic Zones and territorial waters, states may take necessary steps to prevent passage which is not innocent, suspend temporarily the innocent passage of foreign ships (United Nations 1982) (UNCLOS, art. 25 (1) and (3)) or levy charges for specific services (UNCLOS art 26(2)). Beyond the territorial waters, at the high seas, the ability of national states to control, to monitor, and to govern environmental, spatial, social and economic processes at sea has diminished.

Each of the Arctic shipping governance arrangements institutionalizes in a specific way and this chapter will discuss the type of liquid institutionalization, in terms of institutional change, reflexivity and power. Section 3.2 presents the characteristics of Arctic shipping, such as the types of Arctic navigation and a description of possible future shipping routes. Section 3.3 starts by describing the institutional setting of Arctic governance, more specifically Arctic shipping, followed by analysis of the emerging Arctic shipping governance arrangements for each of the three shipping routes. In Sect. 3.4 the main question will be answered as to what type of liquid institutionalization is characteristic for Arctic shipping.

3.2   CHARACTERISTICS OF ARCTIC SHIPPING; TYPES AND ROUTES[4]

This section presents the characteristics of Arctic shipping, such as the accessibility of navigation in the Arctic region, the different shipping routes, which are possible in the future with diminishing sea ice covering, and types of shipping.

Because the extent and volume of Arctic sea ice are diminishing (Keil 2018) (see Fig. 3.1), four Arctic shipping routes are emerging (see Figs. 3.2 and 3.3), giving possibilities for different forms of navigation in

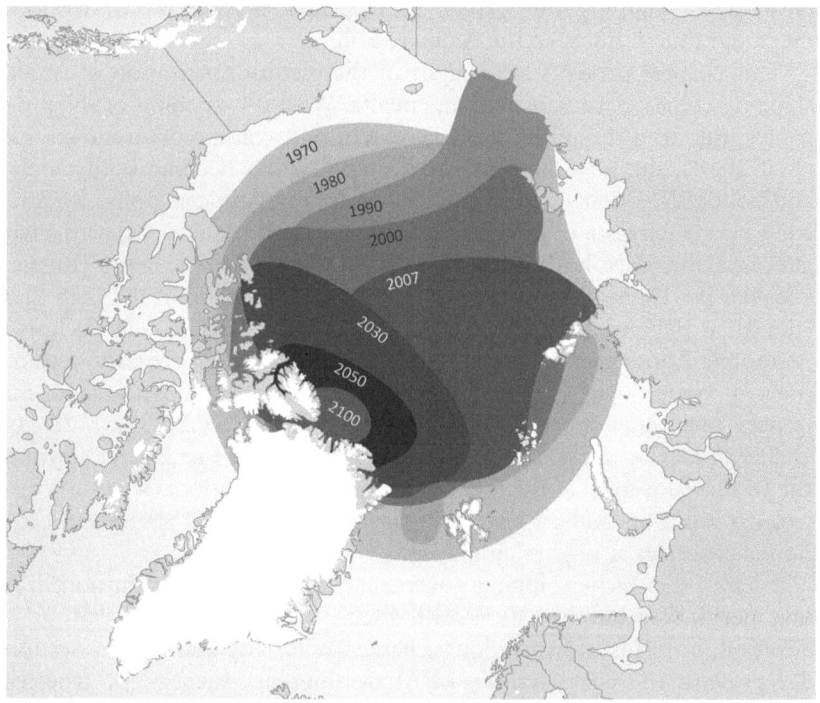

Fig. 3.1 Diminishing Arctic sea ice (1970–2100). (Source: Humpert and Raspotnik 2012)

[4] Parts of Sects. 3.2 and 3.3 are based on Van Tatenhove (forthcoming, Anna-Katharina Hornidge, Maria Hadjimichael (eds.) Ocean Governance. Pasts, Presents, Futures, in MARE Springer book series).

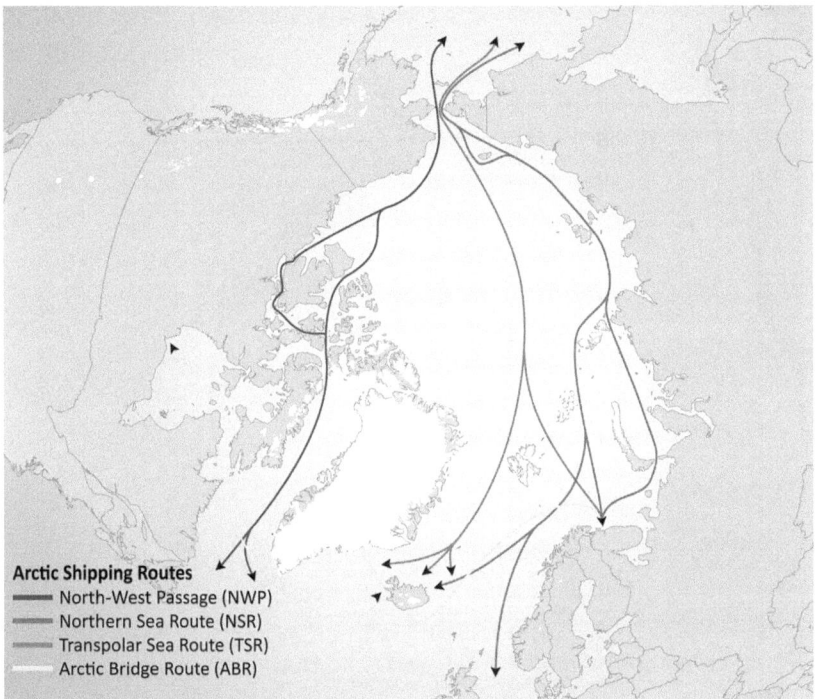

Arctic Shipping Routes
- North-West Passage (NWP)
- Northern Sea Route (NSR)
- Transpolar Sea Route (TSR)
- Arctic Bridge Route (ABR)

Fig. 3.2 Arctic shipping routes. (Source: Humpert and Raspotnik 2012)

the Arctic region,[5] such as liner shipping, [6] bulk shipping (liquid and dry), specialized shipping (LNG (Liquefied/Liquid Natural Gas) and reefer[7]) and cruise shipping.

The Northwest Passage (NWP) connects the Atlantic Ocean with the Pacific Ocean via Canada's Arctic Archipelago. The Northeast Passage (NEP) connects the Atlantic Ocean also with the Pacific Ocean, but from Northwest Europe around the North Cape and along the coasts of Eurasia and Siberia through the Bering Strait. Part of the NEP is the Northern Sea

---

[5] www.worldshipping.org; visited 15/01/2020.

[6] Liner shipping is the service of transporting goods by means of high capacity, ocean-going ships that transit regular routes on fixed schedules (e.g. containerships and roll-on/roll-off ships).

[7] Reefer is a specialized ship to carry frozen products, such as fish and meat (https://pame.is/index.php/projects/arctic-marine-shipping/, visited 17 June 2021).

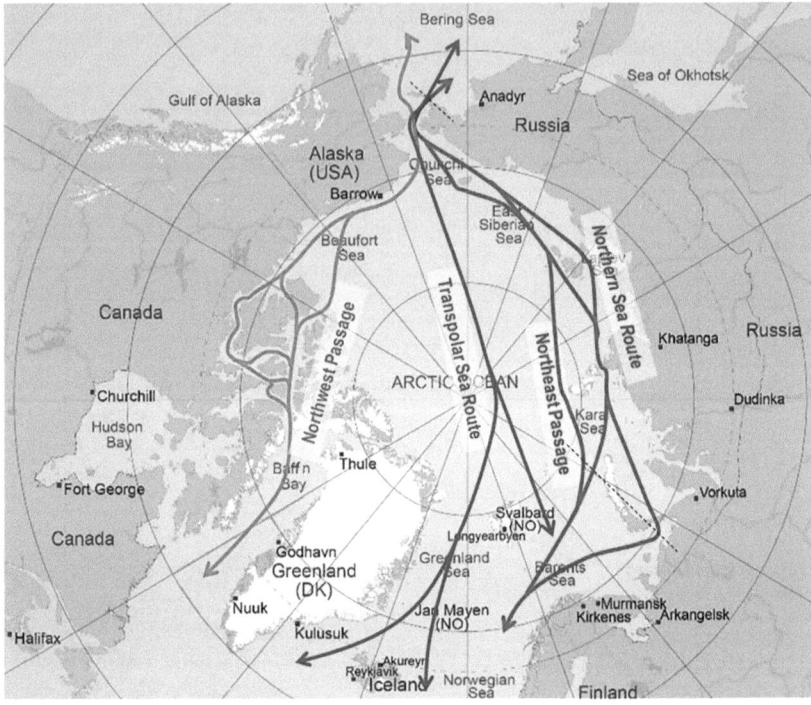

Fig. 3.3 Arctic shipping routes. (Source: Dyrcz 2017)

Route (NSR), which runs from Kara Strait (a small passage between Russia and Novaya Zemlya) to the Bering Strait. The main difference between the NSR and the NEP is that the latter comprises the Barents Sea and provides access to the port of Murmansk (Buixadé Farré et al. 2014). In contrast to the NWP and the NEP, the Transpolar Sea Route (TSR) or Trans-Arctic route does not run through territorial seas or Exclusive Economic Zones of Arctic seas but through Arctic high seas. This route also connects Europe and Asia but is much shorter than the coastal NWP, NEP and NSR routes.

Because the TSR passes outside territorial waters of Arctic states and is shorter in length than the outer Arctic routes, this route is of significant geopolitical and economic importance. Finally the Arctic Bridge Route (linking the port of Murmansk in Russia with the port of Churchill in Canada via Greenland and Iceland) could be a potential future route between Europe and Asia (Humpert and Raspotnik 2012). However, the Arctic Bridge Route will not be discussed in this chapter, because it does not connect the Pacific Ocean with the Atlantic Ocean.

Despite discussions about opening up the Arctic for navigation and other maritime activities, shipping as an activity will remain a challenging activity in the immediate future, due to harsh weather conditions, free floating sea ice, remoteness, lack of communication and of SAR (Search and Rescue) capabilities (Buixadé Farré et al. 2014; Humpert and Raspotnik 2012; Dyrcz 2017). Humpert and Raspotnik (2012) estimated that during the summer season (July–September) the maritime accessibility of the Arctic will increase (see Table 3.1). The ice-free period along the Arctic's main shipping routes is expected to increase from 30 days (2010) to more than 120 days (2050). "However, free-floating ice in summer will remain a serious threat to navigation, and widespread ice in winter will continue to obstruct passage by most ships" (Buixadé Farré et al. 2014: 321).

The harsh conditions in the Arctic Ocean require technical innovations, referred to as "winterization" by Buixadé Farré et al. (2014: 313): which address the challenges unique to sub-zero environments (e.g. icing, snow, rain and fog). For example, winterization solutions include building structures resistant to low temperatures, anti-freezing measures, the procurement of freezing-resistant supplies, etc.

In general, to connect the Atlantic Ocean with the Pacific Ocean, the Arctic shipping routes (NWP and NEP) are shorter than the Suez Canal/ Strait of Malacca and Panama Canal routes, with related economic advantages. To navigate from Rotterdam (the Netherlands) to Shanghai (China)

Table 3.1 Maritime accessibility in 2000–2014 and 2045–2059 for Type A vessels (light icebreaker) in the period July–September

Route	Length (km)	% accessible 2000–2014	% accessible 2045–2059	Accessibility change (%) relative baseline
Northwest Passage	9324	63	82	30
Northern Sea Route	5169	86	100	16
Transpolar Route	6960	64	100	56
Arctic Bridge	7135	100	100	0

Source: Humpert and Raspotnik (2012: 288)

is 25,588 km (via the Panama Canal), 19,550 (via Suez Canal/Strait of Malacca), 17,570 km (via NWP) and 15,793 (via NEP) (Østreng et al. 2013) (pp. 50). However, Arctic shipping will not serve as a substitute for existing shipping routes, rather it will supplement them and provide additional capacity.

3.3   Emerging Arctic Shipping Governance Arrangements

The aim of this section is to reconstruct emerging Arctic shipping governance arrangements of the three main Arctic shipping routes; the Northeast Passage NEP (including the Northern Sea Route, NSR), the Northwest Passage (NWP) and the Transpolar Sea Route (TSR). For each Arctic route, a shipping governance arrangement is constructed, consisting of (coalitions of) public and private actors (governments, international organizations, harbour authorities, shipping sector, coastal communities, and NGOs), discourses, resources and rules. Section 3.3.1 starts with a description of the institutional setting of Arctic governance. The institutional setting of Arctic governance and processes of political modernization are the structural and cultural conditions in which the institutionalization of governance arrangements takes place. The specific development of the dimensions of Arctic shipping governance arrangements related to the three Arctic shipping routes will be analysed in Sects. 3.3.2, 3.3.3, and 3.3.4.

3.3.1   The Institutional Setting of Arctic Governance

The contemporary institutional setting of the Arctic is highly complicated, politicized and fragmented. It consists of formal and informal institutions each with their own responsibilities and interests, such as the Arctic states organized as the Arctic Five,[8] and represented in the Arctic Council, the United Nations (represented by the International Maritime Organization (IMO) and the Commission on the Limits of the Continental Shelf (CLCS)), the Northern Dimension (an Intergovernmental Platform of

---

[8] The Arctic five are the five Arctic littoral states: Russia, Norway, Denmark (Greenland), Canada, and the USA (Alaska).

cooperation between the EU, Russia, Norway, and Iceland[9]) and the Nordic Council (an inter-parliamentary coalition between the Nordic countries[10]). In Sect. 3.3.1.1 the main Arctic shipping governance institutions will be presented, followed by a characterization of Arctic structural and cultural conditions, based on the state-led governance architecture Sect. 3.3.1.2.

Main Institutions of Arctic Shipping Governance
The **Arctic Council** (AC), established in 1996 with the Ottawa Declaration, is an intergovernmental forum to promote cooperation, coordination and interaction among the Arctic states. What makes the Arctic Council special is the formal membership of representatives of Arctic Indigenous communities and other Arctic inhabitants. Arctic states and Indigenous communities discuss and decide on common Arctic issues, in particular issues of sustainable development and environmental protection in the Arctic. The Arctic Council has decision-making power in which also non-Arctic states want to participate (Koivurova 2013; Smits et al. 2014; Koivurova 2009). The Council consists of the eight Arctic states[11] and six organizations representing Arctic Indigenous people.[12] Observer status in the AC is open to non-Arctic states, intergovernmental and interparliamentary organizations with a global and/or regional constituency, and NGOs that the Council determines as potential contributors to its work.[13] The primary role of observers is to observe the work of the Arctic

---

[9] The Northern Dimension was launched in 1997 by Finland to emphasize the interdependence between the EU and Russia, Norway, Iceland, and the Baltic States (non-EU member states at that time).

[10] Members of the Nordic Council are Denmark, Finland, Iceland, Norway, Sweden, the Faroe Islands, Greenland, and Åland. Parliamentarians of all Nordic countries are taking place in the Council and decide upon issues after which they call on the governments of the Nordic countries to implement these.

[11] Canada, Denmark (including Greenland and the Faroer Islands), Finland, Iceland, Norway, the Russian Federation, Sweden and the USA.

[12] the Arctic Athabaskan Council, Aleut International Association, Gwich'in Council International, Inuit Circumpolar Council, Russian Association of Indigenous Peoples of the North, and Saami Council (https://arctic-council.org/index.php/en/about-us/permanent-participants, visited 17/06/2021).

[13] See (https://arctic-council.org/index.php/en/about-us/arctic-council/observers visited 17/06/2021) for the list of observers.

Council, and to contribute to the work of one of the six Working Groups[14] of the Arctic Council. The AC is increasingly an "active regional organization" (Buixadé Farré et al. 2014). An example is the Arctic Marine Shipping Assessment (AMSA) of 2009 (Arctic Council 2009), which outlined recommendations resulting in the first two binding circumpolar treaties: Cooperation on Aeronautical and Maritime Search and Rescue in the Arctic (2011) and the Agreement on Cooperation on Marine Oil Pollution Preparedness and Response (2013).

The sub-regional institution **Barents Euro-Arctic Council**, established in 1993 is the intergovernmental forum for cooperation for the Barents Region. The Steering Committee for the Barents Euro-Arctic Transport Area (BEATA), focuses on improving the integration of road, railway and port systems in the region, including development of coastal shipping and sea safety Stokke 2013). Similar to the Arctic Council, the Barents Euro-Arctic Council uses knowledge generation and soft law as the main instruments in achieving cooperation (van Leeuwen 2016: 133).

The **United Nations Convention on the Law of the Sea** (UNCLOS) is the international legal framework regulating human uses of sea, by enshrining freedom of navigation, specifying rules to territorial claims, and determining economic rights and broad environmental responsibilities (Buixadé Farré et al. 2014). UNCLOS makes a distinction between the territorial sea (12 nm), the Exclusive Economic Zone (200 nm) and the High seas and specifies the role and jurisdiction of coastal states, port states and flag states. Ships of all states have the right of innocent passage (art. 17) through the territorial waters of (Arctic) states. Passage is innocent so long as it is not prejudicial to the peace, "good order" or security of the coastal State (art. 19(1)), and the coastal state may adopt laws and regulations relating to innocent passage through the territorial sea (art. 21(1)). See Arctic Council assessment (Arctic Council 2009).

The UN **International Maritime Organization** (IMO) regulates shipping by setting standards and regulations about safety, security, efficiency and environmental responsibility. Examples of IMO regulations are

---

[14]There are six Working Groups of the Arctic Council: Arctic Contaminants Action Program (ACAP); Arctic Monitoring and Assessment Programme (AMAP); Conservation of Arctic Flora and Fauna (CAFF); Emergency Prevention, Preparedness and Response (EPPR); Protection of the Arctic Marine Environment (PAME); Sustainable Development Working Group (SDWG) (https://arctic-council.org/index.php/en/about-us/working-groups visited 17/06/2021).

the *International Convention on the Prevention of Pollution from Vessels* (MARPOL), the *International Convention on the Safety of Life at Sea* (SOLAS), *International Convention on the Control of Harmful Anti-Fouling Systems on Ships* (Anti-fouling Convention), *International Convention for the Control and Management of Ships' Ballast Water and Sediments* (Ballast Water Management), *International Convention on Oil Pollution Preparedness, Response and Co-operation* (OPRC). To improve the safety of shipping in the Arctic and to reduce the impact of shipping on the environment IMO's International Maritime Safety Committee established in July 2014 the *International Code for Ships Operating in Polar Waters* (The Polar Code).[15] The Polar Code covers all shipping related matters in Arctic and Antarctic waters, ranging from ship design, construction and equipment, operational and training concerns, "search and rescue" to the protection of the environment and eco-systems of the Polar Regions.[16] The rationale of the Polar Code is that sustainable Arctic shipping is based on two pillars; human safety and environmental protection (Keil 2018). The environmental pillar of the Code consists of binding requirements and regulations relating to oil, invasive species, sewage, garbage, and chemicals and defines three categories of ships.[17] The ship safety pillar formulates binding requirements and regulations concerning equipment, design and construction, operations and manning with the aim "to provide for safe ship operation and the protection of the Polar environment by addressing risks present in Polar waters and not adequately mitigated by other instruments."[18] The implementation and the enforcement of the Polar Code will have implications for a diversity of actors, such as ship-owners, assurance companies, trainers, operators, surveillance and controlling agencies, etc. The Polar Code defines a new stage of Arctic shipping, because it will both constrain navigational operations in the Arctic through binding requirements, while at the same time, favouring a discourse of sustained Arctic shipping, by stimulating and enabling shipping activities, as it contributes to shape the necessary information,

---

[15] The Polar Code is developed as a complement to existing documents, as the new 14th Chapter of the SOLAS Convention and entered into force 01/01/2017.

[16] http://www.imo.org/en/MediaCentre/HotTopics/polar/Pages/default.aspx visited 17/06/2021.

[17] https://www.imo.org/en/MediaCentre/HotTopics/Pages/Polar-default.aspx visited 17/06/2021.

[18] https://www.imo.org/en/MediaCentre/HotTopics/Pages/Polar-default.aspx visited 17/06/2021.

communication and material infrastructures that support shipping activities. According to Keil (2018: 46), the Polar Code does not conclude, that "Arctic shipping is too dangerous or risky (…) and should therefore not take place," but it is an expression of a dominant discourse that "Arctic shipping is seen universally as an activity that can be conducted sustainably (…)." "Arctic shipping is considered to be capable of interacting with the natural environment, Arctic communities, and business interests in a way that enables these assets to co-exist over time without threatening the existence of nature, societies or businesses; thus, their relationship is regarded as fundamentally sustainable."

The **Port State Control** is the result of the aftermath of the enormous oil spill accident with the oil-tanker Amoco Cadiz in 1978. To find solutions for the enforcement gap of IMO standards by ports of 14 European states signed the Paris Memoranda of Understanding (MoU) on Port State Control in 1982. "Under this MoU port states agreed to inspect 25 per cent of ships visiting their ports. After inspection a port state can request the rectification of a deficiency, detain or ban a ship. The outcomes of the inspections are registered in a database" (van Leeuwen 2016: 136). The MoU has developed a targeting policy using a ship's risk profile to base the scope, frequency and priority of inspections. The MoU also requires performance lists of individual ships, classification societies and flag states. "While no separate MoU on Port State Control exists for the Arctic, all Arctic states except the US participate in the Paris MoU. The US does not participate in any of the MoUs but ensures inspection of visiting ships through its Coast Guard, sharing most of the requirements of the MoUs" (van Leeuwen 2016: 136).

The Institutional Governance Setting of Arctic Shipping
According to Van Leeuwen (2016: 136) Arctic shipping governance is an example of state-led governance. The Arctic Council, IMO, MoU on Port State Control are based on inter-state cooperation and decision-making. Van Leeuwen gives the following differences between the authority of Arctic shipping governance institutions. Arctic institutions, such as the AC and sub-regional Arctic institutions, do not develop rules, but guidelines, knowledge generation and norms as a base for future collaboration. In contrast, IMO has a strong standard setting authority in Arctic shipping governance, reflected in the safety and environmental rules related to shipping in the Polar Code, but IMO lacks the authority to generate compliance, which is the authority of flag states (delegated to classification

societies) and port states (within the setting of MoUs on Port State Control). Van Leeuwen (2016: 137), building on Humrich (2013) concludes that "the architecture of Arctic governance exhibits cooperative, rather than competitive, fragmentation, because the Arctic Council explicitly builds on norm generation in other institutions," while "private authority is of limited relevance to Arctic shipping governance. There is, nonetheless, an element of delegated authority through the classification societies who are tasked with the monitoring of IMO standards, especially when it comes to standards that require a certificate to show compliance."

This fragmented intergovernmental institutional governance setting defines the conditions of cultural and structural conditioning in which Arctic shipping governance arrangements institutionalize.

There is one dominant discourse related to Arctic navigation that frames shipping in all three Arctic shipping governance arrangements as a legitimate activity in the Arctic region, with the accompanying storyline that navigation in Arctic waters can be sustainable under certain circumstances and conditions. The rule supporting this storyline is an effective implementation and enforcement of the Polar Code. This code is a crucial condition for sustainable Arctic shipping, because it addresses risks present in polar waters, which are not adequately mitigated by other instruments. The analysis will give insight in different types of Arctic shipping governance arrangements and forms the basis to analyse the processes of liquid institutionalization of Arctic shipping (Sect. 3.4).

3.3.2 The Northeast Passage (NEP) and the Northern Sea Route (NSR) Shipping Governance Arrangements

According to Buixadé Farré et al. (2014), the NEP is the most practicable route in the Arctic both as a transport corridor for natural resources and as a shorter route for transit shipping. Although this Arctic route has the highest potential for transit shipping and transporting resources, there will be serious challenges for container shipping, because they operate under a just-in-time regime, which relies on predictability and precise schedules. Bulk cargo ships do not require such a regime; therefore, it is more likely that bulk-cargo ships can deal with the variability of the NEP. However, despite potential for bulk resource transport there remain significant physical and logistic limitations, such as shallow bathymetry (Arctic Council

2009). The shallow depths of the NEP/NSR make it impossible for the new generation ultra large container ships (ULCS) to transit, meaning these ships will continue to prefer the Strait of Malacca/Suez Canal Route.

Tankers, general cargo shipping and icebreakers perform the navigation activities at the NEP/NSR route. Of the 207 transits (between 2011 and 2015), 45% were tankers and 17% general cargo.[19] During the winters (January–April 2017–2019), shipping activities took place mainly west of the Kara Sea.[20]

Within the NSR shipping governance arrangement Russia has a dominant position. Although most Russian regulations are consistent with international law and requirements (UNCLOS and IMO), the country has adopted rules "pertaining to vessels operating in the NSR that contain certain provisions that go beyond international rules and standards (for example, inspections, requirements for ice pilots and transit fees)" (AC 2009: 119). This is a reflection of the Russian Arctic Strategy in which Russia sees the utilization of the NSR as a national integrated transport and communication system to safeguard Russian interest in the Arctic (Buixadé Farré et al. 2014: 308). Russia perceives the NSR as a national waterway (AC 2009) and has developed a framework that obligates all ships to request permission to access the NSR, and to deny passage for political reasons (e.g. in 2013, Russia denied the requests made by Greenpeace's icebreaker Arctic Sunrise to enter the NSR three times).

Another important resource for the future viability of the NEP/NSR is the availability and accessibility of ports. The current availability of Russian ports for repairs and maintenance is scare. In 2014, it was reported that of the 18 marine ports in the Russian Arctic, 11 are in poor condition and located in regions with sparse land transportation infrastructure, only 4 ports (Murmansk, Arkhangelsk, Vitino and Kandalaksha) are in fairly good condition (Buixadé Farré et al. 2014) (313) (see Fig. 3.4).

Liu et al. (2021), studied the effect of Russia's Arctic Strategy on the development of ports along the NSR. "The Foundations of Russian Federation Policy in the Arctic until 2020 and Beyond" is according to Liu et al. (2021: 4) the first general "outline of Russia's Arctic

---

[19] https://pame.is/index.php/projects/arctic-marine-shipping visited 17/06/2021.

[20] In 2018: 278 transits (124 tankers; 34 LNG tankers; 59 icebreakers; 34 containerships; 26 general cargo and 1 SAR). In 2019: 426 transits (144 tankers; 118 LNG tankers; 86 icebreakers; 65 containerships; 1 bulk and 1 SAR) (Centre for High North Logistics Nord University, 2019).

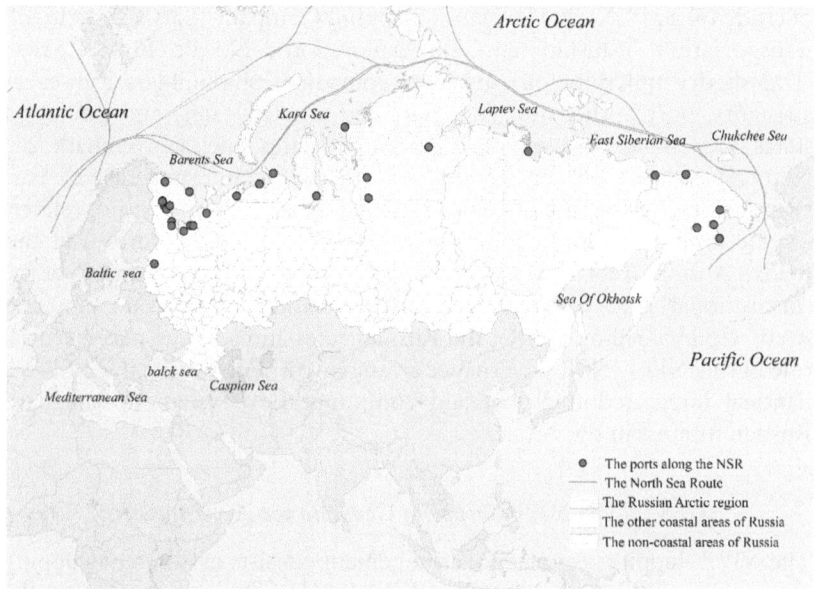

Fig. 3.4 Ports along the NSR in (Source: Liu et al. 2021)

development program for the 21st century, which has elaborated on the implementation measures of Russia's Arctic Strategy and is an important basis for Russia to formulate the Arctic policy since then." Their analysis showed that the implementation of the Arctic Strategy has not promoted the cargo throughput of ports along the NSR during 2003–2012. The authors conjecture that the development of the ports along the NSR is restricted by the low level of economic development and foreign trade, as well as the lag of transportation infrastructure in the Russian Arctic region.

Russia is responsible for the coordination of "Search and Rescue" (SAR) activities along the NSR. Although Russia has invested in the creation of 10 SAR centres along the NSR, substantial parts of the NSR lie outside the coverage of these centres, making Russian icebreakers the only potential respondents to a SAR request (Farré et al. 2014: 314–315).

The NEP/NSR governance arrangement consists of coalitions and infrastructures related to tankers (oil and LNG) and general cargo shipping. The main governmental and non-governmental actors of this coalitions are shipping companies, such as Sovcomflot (Russian, transportation of crude oil and LNG), Murmansk Shipping Company (partly Russian, oil transportation, transhipment and exploration), Nordic Bulk Carriers (Danish, dry bulk shipping), insurance companies, shipbuilders, icebreaker assistants, port authorities, flag, port and coastal states, and interested states like China, Norway and EU member states, such as Denmark, the Netherlands and Germany. Examples of the needed infrastructure are harbour facilities (repair, maintenance, storage, processing industries, refineries, etc.), search and rescue facilities, hinterland infrastructures (rail and roads). Although the institutional setting of the NEP/NSR consists of the (institutional) rules of UNCLOS, IMO regulations (Polar code, etc.) and Arctic Council ruling (SAR), the Russian rules and strategy play a crucial role in this NEO/NSR governance arrangement. For Russia, the NSR is a national integrated transport and communication system to safeguard Russian interest in the Arctic.

3.3.3   The NWP Shipping Governance Arrangement

The NWP shipping governance arrangement consists of different shipping regime complexes in the Canadian Arctic, such as "community re-supply; bulk shipments of raw materials, supplies and exploration activity for resource development operations; and tourism" (AC 2009: 113).

The St Roch (a Canadian ship) realized the first complete transit from west to east in 1942, followed by the oil tanker "Manhattan" in 1969. During the period 1969–1990, there were only 30 complete transits. In 2012, 30 vessels transited through the NWP, while in 2014 only 17 vessels managed the full transit. In 2013 for the first time, a large bulk carrier transited the NWP.[21] These figures indicate that it is unlikely that there will be commercial shipping on a regularly basis to transit the NWP from west to east, aside from a few small specialty cruise operators (AC 2009: 114). Apart from this specialized cruise ship tourism, it is not expected that the NWP will be a viable trans-Arctic route in the nearby future, "due to seasonality, ice conditions, a complex archipelago, draft restrictions, chokepoints, lack of adequate charts, insurance limitations and other costs" (AC 2009: 114). However, according to the Arctic Council there will be an increase in destinational shipping in the Canadian Arctic driven by

[21] https://www.enr.gov.nt.ca/en/state-environment/73-trends-shipping-northwest-passage-and-beaufort-sea (visited 17/06/2021).

increasing demand for seasonal re-supply activity, expanding resource development and tourism (AC 2009: 114).

Former inaccessible destinations such as the North Pole, the Northwest Passage and the Northern Sea Route are more and more open for the public. This has resulted in an increase of Arctic cruise tourism; between 1984 and 2004, 23 commercial cruise ships accomplished transits of the Northwest Passage, while in 2008 seven commercial tours were planned. The Arctic tourism industry ranges from relatively small expedition style vessels that hold less than 200 people, to large luxury cruise liners that can hold 1000 or more. According to Cajaiba-Santana et al. (2020) cruise ship tourism in the Arctic is based on the "expedition" model of Arctic cruising (Cajaiba-Santana et al. 2020), involving small vessels (between 20 and 500 passengers). Expedition cruise tourism is about "shore landings and exploration using rubber boats, quality environmental and historical interpretation of biodiversity, landscapes, historical remains and current use, remote and exclusive wilderness experience, minimal environmental and social impact, human safety and flexibility depending on dynamic weather and sea-ice conditions" (van Bets et al. 2017) (p. 1585).

Most of the passenger vessel traffic takes place along the Norwegian coast, the coasts of Greenland,[22] Iceland and Svalbard. Though there is some passenger vessel traffic in the Canadian Arctic and Alaska, those numbers were small in comparison to the higher traffic areas. Important destinations in the NWP organized by Polar Cruises[23] are Spitsbergen (Svalbard), from Kangerlussuaq (Canada) to Nome (Alaska) and from Greenland to the Bering Sea, the west and east coasts of Greenland (west and east), and Baffin Island.[24] In general, cruise tourism in the NWP consists of cruise tour operators, cruise-owners, expedition leaders, the AECO (Association of Arctic Expedition Cruise Operators), shipbuilders, Arctic and non-Arctic states, consumers, scientists. According to Cajaiba-Santana et al. (2020), cruise shipping in the Arctic is facing several institutional

---

[22] Cruise ship traffic off the coast of Greenland is increasing rapidly. Between 2006 and 2007, port calls into Greenland increased from 157 to 222 cruise ships. The number of port calls in 2006 combined for a total of 22,051 passengers, this increased to a total of 122,314 passengers for all Greenland's harbours in 2019 (http://bank.stat.gl visited 17/06/2020) more than doubling Greenland's total 2021 population of 56,421.

[23] https://www.polarcruises.com/arctic (visited 14/02/2020).

[24] "Polar cruises" has also cruises in the NEP: from Norway to Alaska, from Nome (Alaska to Murmansk), and from Tromsø to Nome (Alaska); Iceland; Newfoundland and Labrador; Russian Far East and Scotland/Ireland.

voids, such as a lack of central authority governing the sector, a lack of regulatory power by AECO, inconsistencies related to the multi-jurisdictional and transnational operating context, and gaps related to for example licencing and Polar Code training requirements, the lack of models for insurance and assessment, and the chartering of uncharted waters.

3.3.4   The TSR Shipping Governance Arrangement

The Transpolar Sea Route is a mid-ocean route and is shorter than the NWP and NEP. Because the TSR has a multitude of possible navigational routes, it is more interesting for bulk shipping (which follows less predictable schedules) than for liner shipping (which is dependent on regular routes and fixed schedules). According to (Humpert and Raspotnik 2012: 294) the challenge for Arctic shipping is not primarily technological, but economic. The lack of schedule reliability and variable transit time along the Arctic shipping routes is a major obstacle for the development of the TSR. In addition, navigation at the TSR remains an unviable option in the near future due to climate conditions and economic uncertainties. To become economically profitable, a different kind of economic optimization needs to be developed, taking into account "the lack of economic hubs, the cost associated with different types of Arctic shipping and uncertainties with regard to investments for special equipment and insurance" (Humpert and Raspotnik 2012: 301).

Compared to the other Arctic shipping routes, the TSR involves only limited legal uncertainties and controversies, because it lies outside the EEZs of Arctic states and is therefore subject to UNCLOS and to High Seas regulations. From a geopolitical perspective the role of China will be crucial. It is expected that China will dominate future global trade, and navigations routes, while China perceives itself (pp. 298) as a "near-Arctic state" and as "a stakeholder." China prefers the TSR to avoid Russian territorial waters and wants to establish a strategic partnership with Iceland. Because Iceland is strategically located in the Northern Atlantic, it may become an important trans-Arctic shipment hub.

Although the TSR is mainly a potential route, only navigated by icebreakers, it is expected that this route could become the dominant Arctic route for bulk shipping in the second half of the twenty-first century. A situation China is anticipating upon by investing in Iceland and by establishing free trade negotiations between China and Iceland in 2009 (Stanley 2012 as cited by Humpert and Raspotnik 2012: 289) to guarantee an important hub in this route.

3.4 Liquid Institutionalization, Power and Reflexivity in Arctic Shipping

The Arctic shipping governance arrangements, related to different (possible) Arctic shipping routes, show different forms of institutionalization, power dynamics and forms of reflexivity. What type of liquid institutionalization is characteristic for the Arctic shipping governance arrangements?

An important trigger of the institutionalization of Arctic shipping is the framing of shipping as a legitimate activity in the Arctic under the conditions of sustainable shipping. Although this discursive space is challenged by some NGOs (e.g. Extinction Rebellion and Ecohustler),[25] the representatives of the shipping sector and governmental actors embrace this dominant discourse. The differences in the institutionalization of the three Arctic shipping governance arrangements are the result of the different roles states play. In the NEP/NSR shipping governance arrangement Russia is able to operate and develop activities beyond international regulations and to define specific national rules for the NSR. As described above, another crucial player is China. The development of the NSR and the TSR is dependent on the economic role of China and the preferred routes by the Chinese government, ship-owners and investors. As it looks now China prefers the TSR to be less dependent of Russia. Both cases are examples of institutionalization as a form of *structural reflectiveness*; the dominant discursive space of Arctic shipping is not challenged, but core actors, such as China and Russia, are able to use rules from different institutional settings to strengthen their position, without changing the Arctic institutional rules.

Arctic shipping is at the beginning of its development. Depending on future ice and weather conditions some shipping routes will be more realistic than others. This makes the future institutionalization, institutional change, and the type of reflexivity of Arctic shipping governance arrangements difficult to predict. Theoretically, one can state that, the future institutionalization of Arctic shipping governance arrangements is affected by the Arctic governance setting. In this regionalized networked polity (van Tatenhove 2016), states are part of the Arctic network state (consisting of UNCLOS, the Arctic Council, the Commission on the Limits of the Continental Shelf (CLCS), the Arctic Five, Permanent Participants

---

[25] Marianne Brooker in the Ecologist. The journal for the post-industrial Age, 21st February 2020.

and Permanent Observers), which is in continuous interaction with the actors within the shipping regime complexes, NGOs and Arctic (Indigenous) communities. Arctic governance is characterized by institutional ambiguity (van Leeuwen et al. 2012), which gives actors the possibility to negotiate and apply the rules and resources from different institutional settings. What kind of institutional change is related to the institutionalization of each of the shipping routes? In general, it looks like that no new institutions have to be developed. The rules set by IMO, national governments, and the Arctic Council define the institutional setting. Type of institutional change depends for example on the possibility to interpret or redefine the environmental and safety rules of the Polar Code.

The development of possible Arctic shipping routes and its power dynamics depends on who are the programmers and switchers of the Arctic shipping governance arrangements. The institutionalization of the Arctic shipping governance arrangements is affected by different coalitions of actors, able to play the role of programmers and switchers. Firstly, the global field of non-state global market actors, such as the global shipping industry (represented by shipping companies and assurance companies) and the global network of ports, are important programmers within each of the governance arrangements. The shipping industry and ports have the resources to invest in types of ships and ports needed for a specific route. Secondly, the local—global network of non-state (civil society) actors, ranging from local Arctic Indigenous communities to international e-NGOs have in their positioning to powerful market parties (dispositioning power) countervailing power by their ability to name and shame (relational power). Next to these market and civil society coalitions several networks of governmental and state actors, such as international organizations (IMO, International Labour Organization (ILO)), the European Union, nation states (Arctic states, China, South Korea, European states, etc.) and intergovernmental/transnational institutions (Arctic Council, Barents Euro-Arctic Region). Depending on the characteristics of the shipping routes these governmental actors can play the role both of programmer and switcher. For example, IMO and the Arctic Council are programmers in each of the three governance arrangements, but China and Russia play an important role as programmers of the NEP/NSR and TSR shipping governance arrangements. China could also play the role of switcher in the selection of the NEP–TSR shipping routes, depending on their willingness for example to invest in Russian ports and in the development of ships needed for these routes. Besides their role as programmer the Arctic Council and IMO could play an important role as switcher between these governance arrangements developing institutional rules and an Arctic shipping vision for the whole region.

To conclude the (future) liquid institutionalization of Arctic shipping governance arrangements resembles Type II of liquid institutionalization (Gelatinous). The analysis of the three shipping governance arrangements shows elements of both structuration/structural elaboration, and stabilization/maintenance. Governmental and non-governmental actors interact within an institutional context of Arctic shipping, which is in development, depending on the physical conditions of the Arctic. Given the institutional ambiguity of the Arctic governance setting, actors will be able to define new rules or change existing rules/institutions (structural reflectiveness) within the dominant Arctic shipping discourse.

REFERENCES

Arctic Council. 2009. Arctic Marine Shipping Assessment 2009 Report.

Cajaiba-Santana, G., O. Faury, and M. Ramadan. 2020. The Emerging Cruise Shipping Industry in the Arctic: Institutional Pressures and Institutional Voids. *Annals of Tourism Research* 80. https://doi.org/10.1016/j.annals.2019. 102796.

Dyrcz, Czeslaw. 2017. Safety of Navigation in the Arctic. *Scientific Journal of the Polish Naval Academy* 4 (211): 129–146. https://doi. org/10.5604/01.3001.0010.6742.

Farré, Buixadé, Scott R. Albert, Linling Chen Stephenson, Michael Czub, Ying Dai, Denis Demchev, Yaroslav Efimov, et al. 2014. Commercial Arctic Shipping through the Northeast Passage: Routes, Resources, Governance, Technology, and Infrastructure. *Polar Geography* 37 (4): 298–324. https://doi.org/10.108 0/1088937X.2014.965769.

Humpert, Malte, and Andreas Raspotnik. 2012. The Future of Arctic Shipping Along the Transpolar Sea Route. *Arctic Yearbook*, 281–307.

Humrich, Christoph. 2013. Fragmented International Governance of Arctic Offshore Oil: Governance Challenges and Institutional Improvement. *Global Environmental Politics* 13 (3): 79–99. https://doi.org/10.1162/ GLEP_a_00184.

IMO. 2014. Reduction of GHG Emissions from Ships. Third IMO GHG Study 2014 – Final Report.

Keil, Kathrin. 2018. Sustainability Understandings of Arctic Shipping. In *The Politics of Sustainability in the Arctic: Reconfiguring Identity, Space, and Time*, ed. Ulrich Pram Gad and Jeppe Strandsbjerg, 34–51. Routledge.

Koivurova, T. 2009. International Governance and Regulation of the Marine Arctic. A Proposal for a Legally Binding Instrument. *International Governance and Regulation of the Marine Arctic.*

Koivurova, Timo. 2013. Gaps in International Regulatory Frameworks for the Arctic Ocean. *NATO Science for Peace and Security Series C: Environmental Security* 135: 139–155. https://doi.org/10.1007/978-94-007-4713-5_15.

Liu, Chuan-Ying, Hou-Ming Fan, Xiu-jing Dang, and Xuan Zhang. 2021. The Arctic Policy and Port Development along the Northern Sea Route: Evidence from Russia's Arctic Strategy. *Ocean and Coastal Management* 201 (February): 105422. https://doi.org/10.1016/j.ocecoaman.2020.105422.

Østreng, Willy, Karl Magnus Eger, Brit Fløistad, Arnfinn Jøgensen-Dahl, Lars Lothe, Morten Mejlnder-Larsen, and Tor Wergeland. 2013. *Shipping in Arctic Waters: A Comparison of the Northeast, Northwest and Trans Polar Passages. Shipping in Arctic Waters: A Comparison Of The Northeast, Northwest And Trans Polar Passages.* Berlin, Heidelberg: Springer Berlin Heidelberg. https://doi.org/10.1007/978-3-642-16790-4.

Smits, Coco, Jan P.M. van Tatenhove, and Judith van Leeuwen. 2014. Authority in Arctic Governance: Changing Spheres of Authority in Greenlandic Offshore Oil and Gas Developments. *International Environmental Agreements: Politics, Law and Economics* 14 (4): 329–348. https://doi.org/10.1007/s10784-014-9247-4.

Stokke, Olav Schram. 2013. Regime Interplay in Arctic Shipping Governance: Explaining Regional Niche Selection. *International Environmental Agreements: Politics, Law and Economics* 13 (1): 65–85. https://doi.org/10.1007/s10784-012-9202-1.

Tianming, Gao, Vasilii Erokhin, Aleksandr Arskiy, and Mikail Khudzhatov. 2021. Has the COVID-19 Pandemic Affected Maritime Connectivity? An Estimation for China and the Polar Silk Road Countries. *Sustainability* 13 (6): 3521. https://doi.org/10.3390/su13063521.

United Nations. 1982. *Convention on the Law of the Sea.* United Nations.

van Bets, Linde K.J., Machiel A.J. Lamers, and Jan P.M. van Tatenhove. 2017. Collective Self-Governance in a Marine Community: Expedition Cruise Tourism at Svalbard. *Journal of Sustainable Tourism* 25 (11). https://doi.org/10.1080/09669582.2017.1291653.

van Leeuwen, Judith. 2016. Governing the Arctic in the Era of the Anthropocene. Does Corporate Authority Matter in Arctic Shipping Governance? In *Environmental Politics and Governance in the Anthropocene. Institutions and Legitimacy in a Complex World*, ed. Philipp Pattberg and Fariborz Zelli, 1st ed., 127–144. London: Routledge. https://doi.org/10.4324/9781315697468.

van Leeuwen, Judith, Luc van Hoof, and Jan P.M. van Tatenhove. 2012. Institutional Ambiguity in Implementing the European Union Marine Strategy Framework Directive. *Marine Policy* 36 (3): 636–643. https://doi.org/10.1016/j.marpol.2011.10.007.

van Tatenhove, Jan P.M. 2016. The Environmental State at Sea. *Environmental Politics* 25 (1): 160–179. https://doi.org/10.1080/09644016.2015.1074386
.

CHAPTER 4

Deep Seabed Mining

Abstract This chapter analyses the institutionalization of Deep Seabed Mining (DSM) governance arrangements and starts by discussing the governance context and practice of deep seabed mining, followed by examining the role of the International Seabed Authority, the role of governmental and non-governmental actors in the exploration and exploitation of marine resources from the seabed and the way this is organized. The process of (liquid) institutionalization of DSM is illustrated by the example of the *Clarion Clipperton Zone* (CCZ), in the Pacific Ocean. The analysis showed both *stabilization*, in which the existing historical power position of pioneer firms and sponsoring states is defended against change, supported by the neo-liberal and neo-mercantilist discourses; and, *structuration*, related to the development of the mining code and the development of the Environmental Management Plan for the CCZ. The analysis indicates characteristics of Type II and Type III liquid institutionalization within the example of DSM.

Keywords Deep Seabed Mining; International Seabed Authority; Clarion Clipperton Zone; Sponsoring states; Pioneer firms; Mining Code; The Area; Common Heritage of Mankind; The Enterprise

© The Author(s), under exclusive license to Springer Nature Switzerland AG 2022
J. P. M. van Tatenhove, *Liquid Institutionalization at Sea*,
https://doi.org/10.1007/978-3-031-09771-3_4

4.1   INTRODUCTION

This chapter discusses the institutionalization of the Deep Seabed Mining (DSM) governance arrangements. The main question is what type of liquid institutionalization is characteristic for DSM, in terms of elaboration/structuration—maintenance/stabilization, institutional change and reflexivity. Section 4.2 starts with a general description of DSM as activity and the institutional governance setting in which DSM is developed. Section 4.2.1 presents what is mined for in the deep sea and which techniques are used (Sect. 4.2.1), followed by a description of the objectives and institutions of the UN organization governing deep seabed mining the International Seabed Authority (ISA) (Sect. 4.2.2). Section 4.2.3 describes the conflicting discourses related to DSM. Section 4.2.4 presents the ISA mining code, followed by the environmental principles of this governance arrangement, such as the precautionary principle, equity and benefit sharing (Sect. 4.2.5). Section 4.3 constructs the emerging DSM governance arrangement related to the planned mining activities in the *Clarion Clipperton Zone* (CCZ). Finally, Sect. 4.4 discusses the type of liquid institutionalization in DSM.

4.2   DEEP SEABED MINING AND ITS
INSTITUTIONAL SET-UP

4.2.1   *What is Deep Seabed Mining?*

Deep seabed mining in the deep and the high seas is the exploration and possible exploitation of mining activities and resources. These mining activities focus on the extraction of minerals and energy sources from the seabed in (vulnerable) deep sea marine ecosystems. Mining activities focus on prospecting, exploration and exploitation of Polymetallic Nodules (PMN), Polymetallic Sulphides (PMS) and Cobalt-rich Ferromanganese Crusts (CFC). PMN are composed of Manganese (Mn), Iron (Fe), silicates and hydroxides. PMN are interesting for mining because of the trace metal contents, such as Nickel (Ni), Copper (Cu), Cobalt (Co), Manganese (Mn) and Rare Earth Elements (REE). The size of these nodules varies from micro-nodules of a few millimetres to about 20 cm with an average of 2–8 cm. These nodules are found in the Clarion-Clipperton Fraction Zone, the Indian Ocean and the Pacific Ocean. PMS is an accumulation or "heap" of minerals composed of Sulphur and metals (e.g. iron, zinc, lead, copper, silver and gold) called sulphides. The term "polymetallic" is used to refer to several metals, with heaps sometimes exceeding 100 m in diameter and several tens of metres in height. Known as vents, these are found in the Indian Ocean and the Mid Atlantic Ridge. Finally, CFC grow on hard-rock substrates of volcanic origin and consist of Cobalt, Platinum, REE, Nickel and Manganese, and are found in the Pacific Ocean and South Atlantic Ocean.

Techniques for exploration within deep sea mining are remote sensing, supplemented with the study of samples collected by draglines grabs and box corers dropped to the seabed to collect samples and the use of Remotely Operating Vehicles (ROVs).[1] There are four basic methods of marine mining or recover mineral deposits: scraping them from the surface; excavating them from a hole; tunnelling to a deposit beneath the surface; drilling into the deposit and fluidizing it. No sustained operations have taken place for the commercial recovery of solid minerals in water depths greater than 200 metres.[2]

4.2.2   The International Seabed Authority

A core element of the DSM governance arrangement is the International Seabed Authority (ISA). ISA is an UN agency established under the 1982 United Nations Convention on the Law of the Sea (UNCLOS) and the 1994 Agreement relating to the Implementation of Part XI of the United Nations Convention on the Law of the Sea (1994 Agreement). ISA is based in Kingston, Jamaica and regulates and manages seabed mining at the Areas beyond National Jurisdiction (ABNJ), also referred to as "the Area" (UNCLOS Art. 1, 1(1) and Art. 157(1)). The Area is the seabed and ocean floor and subsoil thereof beyond the limits of national jurisdiction and is unequivocally declared as "the common heritage of mankind" (Art. 136, UNCLOS). Complying with this mandate includes ensuring the effective protection of the marine environment (Art. 145) and designing an equitable sharing system to distribute financial and other economic benefits derived from DSM operations (Art. 140(2)) (Ginzky et al. 2020). The ISA consists of the Assembly, the Council, the Financial Committee and the Legal and Technical Commission.

---

[1] www.isa.org.jm (visited 19/08/2021).
[2] www.isa.org.jm (visited 19/08/2021).

The *Assembly* consists of 168 member states, which are divided into regional groups.[3] As of April 2020, ISA has 92 observers, 30 observer states (including the USA), 32 intergovernmental organizations (such as UN agencies) and 30 non-governmental organizations (including e-NGOs, such as Conservation International, Greenpeace and the World Wildlife Fund (WWF), and research centres and industry alliances like the World Ocean Council) (ISA 2019b).[4] The Assembly establishes the general policies, sets the two-year budget, approves rules, regulations and procedures that the ISA may establish, governing prospecting, exploration and exploitation in the Area (more information is available here www. isa.org).

The *Council* consists of 36 members, elected by the Assembly,[5] and is the executive organ of ISA. It establishes specific policies in conformity with UNCLOS and the general policies set by the Assembly, it supervises and coordinates implementation of the regime established by UNCLOS

[3] ISA regional groups are: African group; Asian-Pacific group; Eastern European group; Latin-American and Caribbean States; Western European and other States (ISA 2019a).

[4] "Observer States may participate in the deliberations of the Assembly, but are not entitled to participate in the taking of decisions. Observers from the United Nations and its specialized agencies, and other intergovernmental organizations may participate in the deliberations of the Assembly if invited by the President on questions within the scope of their competence. NGO Observers may sit at public meetings of the Assembly and, upon invitation from the President and subject to approval by the Assembly, may make oral statements on questions within the scope of their activities. Written statements submitted by NGO observers within the scope of their activities which are relevant to the work of the Assembly should be made available by the Secretariat in the quantities in the languages in which the statements are submitted. All observers of ISA may designate representatives to participate, without the right to vote, in the deliberations of the Council, upon the invitation of the Council, on questions affecting them or within the scope of their activities" (https://www. isa.org.jm/observers, visited 27/01/2021).

[5] Members are elected in five groups. Group A: four States which consume or import more than 2% of minerals to be derived from the Area, including one Eastern European country. Group B: four States from the eight states with the largest investments in activities in the Area. Group C: four States from States, which, are major net exporters of minerals to be derived from the Area, including at least two developing States whose exports of such minerals have a substantial bearing upon their economies. Group D: six developing States, representing special interests (e.g. large populations, land-locked or geographically disadvantaged, island States, major importers of minerals to be derived from the Area, potential producers for such minerals and least developed States). Group E: 18 states elected according to the principle of ensuring an equitable geographical distribution of seats in the Council as a whole, with at least one member from each geographical region (Africa, Asia-Pacific, Eastern Europe, Latin America and the Caribbean and Western Europe and Others).

to promote and regulate exploration for and exploitation of deep-sea minerals by States, corporations and other entities. The Council draws up the terms of contracts, approves contract applications, oversees implementation of the contracts, and establishes environmental and other standards. More specific it approves 15-year plans, exercises control over prospecting, exploration and exploitation in the area.

The *Legal and Technical Commission* (LTC) is responsible for reviewing mining applications, the supervision of exploration or mining activities and the assessment of the environmental impact of such activities.

The *Finance Committee* consists of 15 members elected by the Assembly for a period of five years taking into account equitable geographical distribution amongst regional groups and representation of special interests. The Committee plays a central role in the administration of ISA's financial and budgetary arrangements.

4.2.3   Conflicting Discourses

The way actors define DSM, the knowledge needed for DSM and the legitimacy of DSM activities revolves around two competing discourses: "*Terra-nullius* versus *Common Property*" (Zalik 2018b: 2). "*Terra nullius*" or no-man's land refers to the position that everybody has free access to and can claim the marine resources in the ABNJ. The "*Common Property*" discourse implies that marine resources and the ABNJ are common heritage of humankind and everybody should have equal access to them. The "common property" discourse is the core principle of ISA. According to Willaert (2020b) the designation of the Area and its mineral resources as common heritage of humankind was the result of a gradual process. In his historical overview, Willaert showed how the Maltese Ambassador, Arvid Pardo, presented the principle of common heritage derived from the notion of *res communis*, at the General Assembly of the UN in 1967. The main motivation was that DSM should not only be an activity of industrialized states, but developing states should also be included, to share in a fair distribution of benefits. This is expressed in art. 136 "The Area and its resources are the common heritage of mankind." However, the focus on equity and redistribution was changed with the 1994 implementation agreement with a focus om market principles and voting procedures that weakened the weight of the Global South by preventing majority rule (Zalik 2018b: 2). To realize this common property ideal UNCLOS (art. 153) has defined "*the parallel system.*" This parallel system refers to the principle that in case of polymetallic nodules an application must be sufficiently large and of sufficient value to accommodate two mining operations of "equal estimated commercial value." One part is to be allocated to the applicant and the other is to become a *reserved area*. These reserved areas are set aside for activities by developing states or by ISA through its *Enterprise*, and is managed by the signatory states of UNCLOS via the ISA. The Enterprise is an important resource through which the International Seabed Authority can develop its own mining activities in the Area and by generating its own benefits for mankind.[6] A draft agreement (December 2018) about a joint venture between Poland and the ISA leads to the operationalization of the Enterprise.

Other principles to govern the Area are a ban on appropriation (over any part of the Area or its resources) (art. 137), the equitable sharing of financial and economic benefits deriving from activities in the Area (art. 140), exclusive use for peaceful purposes (art. 141), international cooperation and knowledge dissemination with respect to marine scientific research (art. 143–144) and protection of the environment (art. 145). Art. 145 stipulates that ISA is required to take necessary measures to ensure effective protection for the marine environment from harmful effects which may arise from mining activities in "the Area."[7]

The research activities in the Area are marked by geopolitical tensions between the Global North and South and unequal power positions related to historical claims of private contractors before UNCLOS came into force (for an extensive overview, see Zalik 2018b). The role of sponsoring states is crucial within the DSM regime (Willaert 2020b). Mining companies who want to pursue mining activities can only apply to the ISA via a sponsoring state. A sponsoring state can be either the State of nationality, or the State, which exercises effective control (directly or through their

---

[6] See LOSC (n 1), Articles 153(2)(a) and 170, Annex III Article 3(1)–(2), Annex IV Article 1; 1994 Implementation Agreement (n 4), Annex Section 2 (note 33 in (Willaert 2020b)).

[7] "To this end the Authority shall adopt appropriate rules, regulations and procedures for inter alia: (a) the prevention, reduction and control of pollution and other hazards to the marine environment, including the coastline, and of interference with the ecological balance of the marine environment, particular attention being paid to the need for protection from harmful effects of such activities as drilling, dredging, excavation, disposal of waste, construction and operation or maintenance of installations, pipelines and other devices related to such activities; (b) the protection and conservation of the natural resources of the Area and the prevention of damage to the flora and fauna of the marine environment" (UNCLOS, art. 145).

nationals) (pp. 6–7). The role of sponsoring states (Willaert 2020b: 7–8) is to ensure that sponsored companies respect the terms of their contract and obligations under UNCLOS, cooperate with the ISA to evaluate the environmental impact of their activities, ensure environmental protection, based on the precautionary principle and, in cases of non-compliance, the sponsoring state is also responsible for considering legal proceedings.

4.2.4   The Mining Code: ISA's Core Rule

The "Mining Code" refers to the comprehensive set of rules, regulations and procedures issued by ISA to regulate prospecting, exploration and exploitation activities of marine minerals in the international seabed Area. *Prospecting* is the search for minerals and an estimate of their shape, size, value and distribution (Willaert 2020a), and does not require the approval of the Authority. *Exploration* is the thorough analysis of the resources, testing of the recovery of systems and technical, economic and environmental studies related to their future extraction (Willaert 2020a: 177). The ISA has issued Regulations on Prospecting and Exploration for, Polymetallic Nodules in the Area (adopted on 13 July 2000, updated on 25 July 2013); for Polymetallic Sulphides in the Area (adopted on 7 May 2010) and for Cobalt-Rich Crusts (adopted on 27 July 2012). Exploitation can only take place once an application to explore has been approved, while the exploration contract creates exclusive rights to explore for minerals for a defined period within a particular location (pp. 178). For example, a maximum of 150,000 km² (for polymetallic nodules), 10,000 km² (polymetallic sulphides), and 3000 km² (cobalt-rich ferromanganese crusts) for a maximum of 15 years.

In its 25th session (July 2019, ISBA/25/C/WP.1) the Council discussed the Draft regulations on exploitation of mineral resources in the Area (prepared by the LTC). These Draft exploitation regulations consist of a set of rules for all categories of mineral resources for 30 years. Most important elements of the draft exploitation regulations are the implementation of the precautionary approach, the consideration of plans of work, the role of regional environmental management plans and the inspection mechanism. Concerning the sharing of financial and economic benefits, special emphasis needs to be placed on the envisaged financial model and the operationalization and modalities of the "Enterprise," the organ through which the Authority can develop its own mining activities in the Area (see Sect. 4.2.3).

Together with recommendations by ISA's Legal and Technical Commission (LTC), the complete set of these regulations make up the Mining Code (LTC), providing guidance for contractors, including on the assessment of the environmental impacts of exploration for polymetallic nodules.

4.2.5   Environmental Principles, Equity, Benefit Sharing and Participation

A key challenge for the ISA is to determine relevant governance principles which support an equitable balance between mining activities in the Area for the benefit of mankind and ensuring effective protection for the marine environment (Warner 2020). The potential mining activities in the Area should incorporate the principles of marine environmental protection and equitable benefit sharing. The ISA's environmental management approach, reflected in the Mining Code, included the common heritage of mankind, the precautionary principle, prior Environmental Impact Assessment (EIA), conservation and sustainable use of biodiversity and transparency (Warner 2020).

Environmental protection relates to the prevention and reduction of pollution and the protection and conservation of natural resources. Compared to other industries DSM lacks the knowledge of the marine environments in which it operates, while, in addition, there exists very little information on the potential effects of mining activities (Jones et al. 2019). To reduce the uncertainty of the environmental impacts of deep seabed mining Jones et al. (2019: 175) suggest the following: firstly, reduce uncertainty through baseline data collection, experimentation and monitoring of activities. Secondly, the development and implementation of area-based management tools, such as Environmental Management Plans (EMPs). Thirdly, applying the precautionary approach and adaptive management to DSM (assessing the results of monitoring the management plan, with the intention to learn and use the findings to revise the management plan). Potential environmental risks of deep seabed mining may extend beyond the boundary of a single mining site, while others may result in cumulative impacts. It is important to develop environmental management approaches for DSM at a more strategic (regional) level, for example Regional Environmental Assessment (REA) and Strategic Environmental Assessment (SEA) Jones et al. (2019: 175).

The *precautionary principle* is crucial (and mentioned in the draft exploitation regulations) and should be understood in conjunction with other principles and obligations, such as the application of an ecosystem approach, the obligation to apply best environmental practices and the requirement to integrate best available scientific knowledge (Willaert 2020a; Warner 2020: 179). Although ISA agreed that the precautionary principle should be central to the implementation of the exploitation regulations, the interpretation of the principle varies. While NGOs emphasize that the precautionary principle should transcend its status as a purely procedural mechanism, some countries highlighted the legitimate interests of contractors, the importance of a level playing field and cost-effectiveness (Willaert 2020a: 180). Also according to Zalik (Zalik 2018a: 5) proceeding with seabed mining in the absence of detailed knowledge would be contrary to the precautionary principle. It is clear that the precautionary principle needs further specification and implementation in the context of DSM.

Regional Environmental Management Plans (REMPs) are crucial for spatial planning and environmental impact assessment. "REMPs indicate which areas of the marine environment represent the full range of habitats, biodiversity and ecosystem structure and function within a specific region, and on the basis of these plans, areas of particular environmental interest can be established where extraction of minerals is prohibited" (Willaert 2020a: 182). Given REMP's importance as a key element of ISA's environmental policy, according to Willeart (2020a: 183) they should be mandatory and be finalized before the start of the exploitation phase, by formulating clear guidelines about REMP's development.

For the implementation of environmental principles and protection measures, an *inspection mechanism* should be developed and implemented. However, the proposed inspection mechanism in the Draft Exploitation regulations does not formulate clear criteria when an inspection should take place, how to implement remote monitoring, while the mechanism is interpreted narrowly, "referring only to recoding the date, time, and position of the mining activities" (Willaert 2020a: 183). Related to inspections, there exists a tension between inspectors assigned by sponsoring states and the ISA responsibility for monitoring (and so avoiding diverse national regimes, as expressed by some countries at the 25th Session of the Council meetings in 2019).[8]

---

[8] Summary of the 25th Session of the ISA Council, note 47, p. 7 (in: Willaert 2020a: 184).

Related to the equitable sharing of financial and economic benefits, the draft exploitation regulations distinguish three possible systems related to contractor payment: *an ad valorem royalty system* (fees are paid according to the quantity and market value of each type of metal that is mined); a *profit-sharing system* (where mining companies pay a percentage of the profit that they make) and a *hybrid model*. The ad valorem royalty system appears to be the most transparent and workable model, but a final decision is yet to be made. According to Willaert (2020a: 185) "introducing a pure profit-sharing represents a great risk and could drastically undermine the objective of equitable benefit sharing, as creative accountancy could easily minimize the official profits of mining companies, resulting in less income for the Authority and, therefore, mankind as a whole, whose interests the Authority represents." (…) "Moreover, also lacking is a system for the distribution of collected royalties among member states, which is fundamental to realizing equitable sharing of financial and economic benefits" (Willaert 2020a: 185).

4.3   AN EMERGING DEEP SEABED MINING GOVERNANCE ARRANGEMENT

This section constructs an emerging deep seabed mining governance arrangement. The construction of such a DSM governance arrangement will be done by analysing the planned mining activities in the *Clarion Clipperton Zone* (CCZ). In Sect. 4.4, the institutionalization of this governance arrangement will be analysed to give insight in the type of liquid institutionalization represented in this scenario.

4.3.1   The Development of the Activities in the CCZ

The CCZ is a seabed area of about 6 million km² in ABNJ in the Eastern Central Pacific Ocean (see Fig. 4.1).

The CCZ is considered as a prime location for commercially exploration and exploitation of polymetallic nodules. Although, scientific exploration and prospecting research have been conducted since the 1960s, no commercial exploration has yet taken place (Lodge et al. 2014). As of January 2021, the ISA has granted exploration contracts to 16 contractors in the CCZ for the exploration of Polymetallic Nodules. Seven contracts

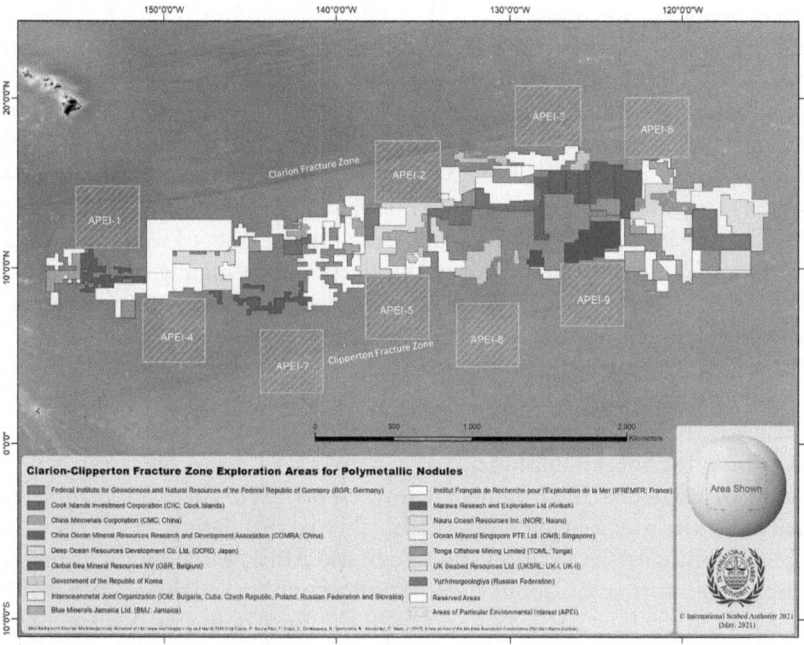

Fig. 4.1 The Clarion Clipperton Fracture Zone. (Source: ISA May 2021, copied from   https://www.isa.org.jm/map/clarion-clipperton-fracture-zone,   visited 12/10/2021)

will expire in 2021 unless they are renewed.[9] Some of these contractors are successor firms to the pioneer investors who had been given exclusive rights to undertake "pioneer activities." For example, Lockheed Martin (via the contractor UK Seabed Resources Ltd.) has an exploration permit for the CCZ,[10] holding only rights under US law. While the USA has only an observer status in ISA (because it has not ratified UNCLOS) it allows, also through agreements with other states, US companies to capitalize on

[9] https://isa.org.jm/files/files/documents/isacont-update.pdf (visited 16/03/2022). March 2022 there is no information if these contracts have been renewed (https://www.isa.org.jm/exploration-contracts/polymetallic-nodules)(visited 16/03/2022).

[10] The USA developed its own legislation on deep sea mining, the 1980 Deep Seabed Minerals Resources Act. Based on this act and because the USA set up agreements with "reciprocating states" (such as the UK, France and Germany) Lockheed got an exploration permit for the CCZ (Zalik 2018b).

their first mover advantage (Zalik 2018b). Moreover, greater access to capital as well as proprietary data keep these companies in pole position. Although UNCLOS promotes information exchange, when it concerns commercially sensitive data, there is much room for exceptions. This especially benefits those with longstanding experience and strong research connections, such as Lockheed Martin that can play the role "as purveyor of data to competing firms" (Zalik 2018a: 347).

The development of the Environmental Management Plan for the CCZ by ISA's LTC is crucial for governing deep seabed mining in this part of the Area. In 2010, the LTC organized a workshop with marine reserve and management specialists to discuss and draft a management plan for the CCZ, which was decided upon in 2011. This plan divided CCZ spatially in three strata/areas for conservation management. The plan included, besides areas for exploration of polymetallic nodules, also the possibility of the establishment of a network of protected areas, called Areas of Particular Environmental Interest (APEI). The plan formulated scientific design principles of APEIs, also incorporating flexibility (the ability to modify the location and size of the APEI, based on for example, improved information about the location of mining activities) (ISA 2011: 6–8).[11] On 26 July 2012, ISA approved the Environmental Management Plan for the CZZ to be implemented in three years, including the designation (on a provisional basis) of a network of APEIs (ISA 2012). Nine APEIs in different bio-geographic sub-regions were designated, with a protected area of 400 Km × 400 Km, placing roughly 25% of the whole CCZ management area under protection. The Council noted in its decision that, by designating APEIs, the precautionary approach as called for by the Regulations had been put in place, and decided that, for a period of five years no application for exploration rights should be granted in these protected areas.

A conflicting issue has been the lack of a comprehensive approach to marine conservation. While ISA exploration contracts require firms to carry out ecological research, including a baseline study, this research does

---

[11] At the 18th session (16–27 July 2012) the Council decided that the "implementation of a comprehensive management plan at the regional level is one of the measures appropriate and necessary to ensure effective protection of the marine environment of that part of the Area known as the Clarion-Clipperton Zone from harmful effects that may arise from activities in the Area and that such a plan should include provision for the establishment of a representative network of areas of particular environmental interest" (ISA 2012, ISBA/18/C/22).

not need to include, or provide, the requisite data needed to meaningfully measure impact (Zalik 2018b). According to the LTC, there is a lack of raw data associated with environment baseline studies, while existing data are also considered to be of insufficient quantity and quality impeding the validity of a regional environmental management plan (Lodge et al. 2014: 69).

4.3.2   The Emergence of a DSM Governance Arrangements in the CCZ

Based on the planned activities in the CCZ, scientific reflections and decisions taken by ISA and other actors within the deep seabed mining regime, this subsection constructs the emerging DSM governance arrangements. The analysis starts with identifying dominant and opposing discourses, and the way changes in discourses result in changes in the other dimensions (i.e. coalitions, rules and resources), resulting in an understanding of the specific institutionalization of DSM governance arrangements.

The main **discourse** of the DSM governance arrangement is the "Common Heritage of Mankind" (CHM). Initially CHM is based on the principle of *res communis*; an area of territory that is not subject to a legal title of any state. The idealistic motivation of CHM was that DSM should not only be an activity of industrialized states and Western oriented mining companies (both firms and states), but that developing states should also be included in the revenues of deep seabed mining. This presupposed the development of a substantive and organizational institutional set-up to implement a system of distribution of equity or equitable sharing of benefits (Ovesen et al. 2018; Bourrel et al. 2018), and assumes the implementation of the following **rules** and **resources**.

Firstly, the implementation of a *parallel system* (a rule to implement CHM), and the development of *the Enterprise* as a commercial mining corporation (the Enterprise as a resource) of the ISA (Sect. 4.3.1). The Enterprise "was initially conceived as a separate organ and empowered by UNCLOS to engage in prospecting and mining in the Area as well as the transporting, processing and marketing of minerals recovered from the Area. When UNCLOS was adopted, the main idea was for the Enterprise to buy the mining technology from commercial operators or to form joint ventures with them. As such, the profits made should be distributed, as part of the CHM, by the ISA. However, the Enterprise has never been set up. With the adoption of the 1994 Implementing Agreement, its role has been revised to include a series of new tasks such as the monitoring of DSM mining trends and developments and the assessment of the results of the conduct of marine scientific research, prospecting and exploration in the Area" (Bourrel et al. 2018). While for the Global South (G-77 States) the development of the Enterprise is crucial to facilitate that mining be carried out under the redistributive principle of CMH (Zalik 2018b: 3, 11), this is not a priority for Northern states. There is no *actual process* for establishing the Enterprise laid out in the strategic plan or draft exploitation regulations (Zalik 2018b: 11). This further expresses the tensions between the global North and global South. Instead of the Enterprise as the organization to organize an equitable distribution of the profits from mining activities, "the balance of power at the ISA favours making the *contracting firm* the body charged with extraction from the spaces designated for profitsharing. (Zalik 2018b: 6). Several joint ventures have been initiated, such as with the Canadian firm Nautilus Minerals in 2013 and a joint venture proposal by Poland in April 2018, which reflected the mercantilist discourse. According to Willaert (Willaert 2020a: 186), there are issues of concern related to this draft proposal (December 2018). These include the absence of any environmental provisions in the draft proposal; the precise financial and liability obligations of the Enterprise; the options regarding dispute settlement; the disclosure of information; the lack of standard rules on joint ventures; and the extent to which this agreement may operate as a precedent for future joint venture agreements. Willaert (2020a: 186) suggests, "to establish a transparent framework for joint ventures with the Enterprise to provide more clarity on the role of the Enterprise during Council and Assembly meetings". (...) "Although several uncertainties remain regarding the precise modalities and costs of the Enterprise, it is crucial that this organ is created swiftly, and therefore, the stipulation in the Draft Exploitation Regulations that the Council shall enable the Enterprise to engage in seabed mining at the same time as other entities is very positive indeed. After all, any failure to operationalize the Enterprise in a timely fashion would affect the direct implementation of the objective of equitable benefit sharing and, therefore, the principle of the common heritage of mankind."

Secondly, the development of *rules to guarantee equal access to (environmental) information and knowledge*. Especially e-NGOs raise the issue of the absence of detailed knowledge (Toonen and van Tatenhove 2020). In 2018, the Brussels-based NGO "Seas at Risks" demanded the Belgian government and ISA to conduct a public consultation regarding the exploration contract of the Belgian company Global Sea Mineral Resources (GSMR), in particular the environmental impact assessment (EIA) related to a mining equipment test (Seas at Risk 2018a). Seas at Risk stated that the procedure was unclear, but succeeded to get their request granted by the Belgian authorities, which organized an open consultation and reviewed GSMR's impact assessment (Seas at Risk 2018b; FPS Economy 2018). Seven comments (of which six Belgian) were received of which two were rejected; the other five were separately addressed. The Belgian government issued an overall response in which it recognized the problem of knowledge gaps, stating "knowledge gained from this test is important in order to set the bar for environmental standards as high as possible" (Belgian Government 2018: 3). It sets out several action points, including the request to GSMR to revise its executive summary of its assessment report. It also recommended to GSMR to make parts of annual reports publicly available, recognizing that the Belgian government itself "is bound, as a member of ISA, to follow the confidentiality rules" (ibid.: 2). The response refers to an announcement made by the LTC chair to set up a workgroup to consider the review process of an environmental impact assessments in the future (ibid.: 3).

Thirdly, rules related to the *environmental management of mining activities*, such as the precautionary principle, Environmental Impact Assessments, polluter pays principle, best available scientific evidence (Warner 2020). The concern of e-NGOs is that proceeding with seabed mining in the absence of detailed knowledge is contrary to the precautionary principle, as the example of "Seas at Risks" showed. In the ABNJ, there is an absence of appropriate systems to govern marine conservation or ensure inter-sectoral collaboration in ecological monitoring. Debates on biodiversity run parallel to (rather than in coordination with) the ISA's development of the extractive regime. "The ISA exploration contracts require firms to carry out ecological research, including baseline research, as part of their ongoing activities. Yet this research does not provide the requisite data to meaningful impact" (Zalik 2018b: 5). The requirement to collect ecological data cultivates the relationship between industry and marine scientists.

Fourthly, the development of *best practices* of transparency and accountability, in which commercial firms (industrial and states) operate under the conditions that the revenues of the exploitation of resources should be shared according to benefit sharing system which is just and fair to developing countries who are not able to employ mining activities themselves.

According to Bourrel et al. (2018) "equitable sharing will only be achieved if there is equitable utilisation of the resources of the Area, understood as requiring a balancing of interests and considerations at stake including for conservation of these resources." However, this CHM discourse was challenged by the economic and geo-political reality as expressed by the "Neo-Mercantilist" and "Neo-Liberalist" discourses. These discourses institutionalized in the ISA governance arrangement with the 1994 implementation agreement and changed the focus of the DSM governance arrangement from equity and redistribution to market principles and voting procedures that weakened the weight of the Global South by preventing majority rule (Zalik 2018b: 2; Bourrel et al. 2018). The balance shifted to coalitions of mining companies and sponsoring states, who claim potentially valuable resources perceived as globally scarce. The mercantilist discourse as a form of economic nationalism, advocates the protection of domestic industries,[12] that is, coalitions of contractors and (national) sponsoring states, and does not advocate helping developing states. According to Zalik (2018b: 2) pioneer firms and states still play a significant role in today's exploitation regulation, because the activities of these industrial actors were protected in the period before the UNCLOS ratification through "pioneer investor" activities agreements. Pioneer firms and states have a historical control over research and technology and the ability to access financial capital. This hampered the development of checks and balances to realize sustainable exploitation and fair benefit sharing regimes. Ardron et al. (2018: 65) (Ardron et al. 2018) identify the following deficiencies related to transparency, unavailability of environmental and safety information; confidential annual reports; unclear quality assurance upon; the compliance of States and contractors to ISA is not reported upon; limited public participation. Another example of best practices of accountability is the *Social License to Operate* (Smits et al. 2017; Voyer and van Leeuwen 2019). Filer and Gabriel 2018 show how SLO poses a particular problem for the operators of deep seabed mining projects because of the uncertainties that surround the definition of the community from whom the license needs to be obtained. "It also shows how different actors, including corporate actors, have tried to shape the "negotiation space" in which to debate the presence or absence of a social licence for the world's first deep sea mine in Papua New Guinea."

---

[12] https://www.thebalance.com/mercantilism-definition-examples-significance-today-4163347 (visited 22/10/2021).

4.4   LIQUID INSTITUTIONALIZATION OF DEEP SEABED MINING

This section addresses the following question: How did the DSM governance arrangement institutionalize and what type of liquid institutionalization is characteristic for DSM?

The practice of DSM and the planning of DSM activities take place in an institutional (cultural and structural) context defined by UNCLOS and specified by the ISA as explained in the sections above. The rules set by the ISA define the exploitation and exploration activities, the responsibilities of mining companies and sponsoring states. While the "Common Heritage of Mankind" (CHM) discourse defines the rules of an equal distribution of the revenues and profits of future mining activities between developed and developing countries, these rules are challenged by the neo-liberal and neo-mercantilist discourses favouring the capitalist status quo. Interactions between the ISA, mining companies, sponsoring states, ISA member states, scientists, NGOs etc. take place and are conditioned by this institutional setting. The development of the Environmental Management Plan and the designation of Areas of Particular Environmental Interest (APEI) in the CCZ showed that actors have room for maneuvering and to negotiate the conditions of environmental management, how to interpret the precautionary principle, the location of APEI after exploitation contracts have been given, etc. According to Toonen and van Tatenhove (2020), *structural reflectiveness* characterizes governance in case of the CCZ. Interactions between actors take place within the structural condition of the neo-liberalist and mercantilist discourses and persistent and longstanding power relations (structural power). These structural conditions and existing power relations position sponsoring states (as market parties) and pioneer firms in a favourable position to ISA and to actors of the Global South. Pioneer firms and sponsoring states use historical rules, and their first mover advantage, based on rights granted before the negotiations and ratification of UNCLOS to manoeuvre within the DSM governance arrangement. While this can be considered a rather morphostatic process, the establishment of the Environmental Management Plan including APEIs sets new institutional conditions for future interactions. Furthermore, the open consultation regarding the Global Sea Mineral Resources (GSMR) exploration contract shows the ability of e-NGOs to apply procedures common in a different governance setting, for example, Belgium, to mobilize an ISA member to improve working practices in the arrangement. Although the CHM discourse of ISA challenges the neo-liberal discourse mining companies and sponsoring states have a strong power position to weaken the CHM discourse, undermining the profit and benefit-sharing of the revenues DSM and by that strengthening the negotiation position of these actors.

Planning processes and interactions in the DSM governance arrangement are influenced by structural and dispositional power. Structural power refers to structures of signification, domination and legitimation, reflected in institutional orders, such as modes of discourse, political/economic institutions and legal institutions (Giddens 1984: 31). The structure of signification, reflected in the CHM and market-based mining discourses, sets the boundaries and possibilities of benefit-sharing and equity within the DSM governance arrangement. These discourses determine the unequal distribution of resources and the positioning of actors' *vis-à-vis* each other in the DSM governance arrangement, but also the capacity of actors to realize outcomes by using these resources (relational power). The order of domination by UNCLOS and ISA, not only positions actors in the DSM governance arrangement but also sets the enabling and constraining conditions of actor's capacity to act. As the main programmers of the CHM discourse in the DSM governance arrangement the Council and the LTC have a dispositional power position in which they are capable to define and implement the exploration and exploitation rules and resources, but also to define the role of private actors. At the same time, the mercantilist discourse provides a strong dispositional power position to the coalitions of mining companies, contractors and sponsoring states, and does not advocate developing states. This dispositional power position of these market actors is the result of the political and economic institutional orders before the ratification of UNCLOS and the establishment of ISA.

When looking at institutional change the DSM governance arrangement shows forms of displacement, drift and layering. On a general level, the establishment of the ISA is an example of defining new rules for an institution in the making, more specific are the mining code and the development of the Environmental Management Plan for the CCZ examples of developing new rules in the arrangement. I label this change as displacement, although it is not about displacing existing rules, but the development of new institutions and institutional rules. Once defined, rules from the neo-liberal and mercantilist discourses, such as a strong position for sponsoring states and pioneer firms are layered upon existing ISA rules.

The non-implementation of the parallel system and the non-development of the Enterprise is an example of institutional change as drift. As defined in Chap. 2, drift is "the failure of relevant decisions makers to update formal rules when shifting circumstances change the social effects of those rules in ways that are recognized by at least some political actors" (Hacker et al. 2015: 184). The ISA is not able to update or implement these institutions, due to the strategies of sponsoring states and contracting firms.

To conclude, the institutionalization of the DSM governance arrangement in the CCZ has most characteristics of Type II liquid institutionalization (Gelatinous). The planning of DSM activities and the development of the DSM governance arrangement show a complex interplay of stabilization/maintenance and structuration/elaboration, forms of structural reflectiveness, while institutional changes are characterized by layering, drift and displacement.

REFERENCES

Ardron, Jeff A., Henry A. Ruhl, and Daniel O.B. Jones. 2018. Incorporating Transparency into the Governance of Deep-Seabed Mining in the Area beyond National Jurisdiction. *Marine Policy* 89 (February): 58–66. https://doi.org/10.1016/j.marpol.2017.11.021.

Belgian Government. 2018. Belgian Response to the Public Consultation.

Bourrel, Marie, Torsten Thiele, and Duncan Currie. 2018. The Common of Heritage of Mankind as a Means to Assess and Advance Equity in Deep Sea Mining. *Marine Policy* 95 (September): 311–316. https://doi.org/10.1016/j.marpol.2016.07.017.

Filer, Colin, and Jennifer Gabriel. 2018. How Could Nautilus Minerals Get a Social Licence to Operate the World's First Deep Sea Mine? *Marine Policy* 95 (October): 394–400. https://doi.org/10.1016/j.marpol.2016.12.001.

FPS Economy. 2018. Environmental Impact Statement from Global Sea Mineral Resources for Small-Scale Testing of Nodule Collector Components on the Seafloor.

Giddens, Anthony. 1984. *The Constitution of Society. Outline of the Theory of Structuration.* Oxford, UK: Polity Press.

Ginzky, H., P. Singh, and T. Markus. 2020. Strengthening the International Seabed Authority's Knowledge-Base: Addressing Uncertainties to Enhance Decision-Making. *Marine Policy.* https://doi.org/10.1016/j.marpol.2020.103823.

Hacker, Jacob S., Paul Pierson, and Kathleen Thelen. 2015. Drift and Conversion: Hidden Faces of Institutional Change. In *Advances in Comparative-Historical Analysis,* ed. James Mahoney and Kathleen Thelen, 180–208. Cambridge: Cambridge University Press. https://doi.org/10.1017/CBO9781316273104.008.

ISA. 2011. Environmental Management Plan for the Clarion-Clipperton Zone (ISBA/17/LTC/7).

———. 2012. Decision of the Council Relating to an Environmental Management Plan for the Clarion-Clipperton Zone (ISBA/18/C/22).

———. 2019a. International Seabed Authority—Authority.

———. 2019b. International Seabed Authority—Observers.

Jones, Daniel O.B., Jennifer M. Durden, Kevin Murphy, Kristina M. Gjerde, Aleksandra Gebicka, Ana Colaço, Telmo Morato, Daphne Cuvelier, and David S.M. Billett. 2019. Existing Environmental Management Approaches Relevant to Deep-Sea Mining. *Marine Policy* 103 (May): 172–181. https://doi.org/10.1016/j.marpol.2019.01.006.

Lodge, Michael, David Johnson, Gwenaëlle Le Gurun, Markus Wengler, Phil Weaver, and Vikki Gunn. 2014. Seabed Mining: International Seabed Authority Environmental Management Plan for the Clarion–Clipperton Zone. A Partnership Approach. *Marine Policy* 49 (November): 66–72. https://doi.org/10.1016/j.marpol.2014.04.006.

Ovesen, Vidar, Ron Hackett, Lee Burns, Peter Mullins, and Scott Roger. 2018. Managing Deep Sea Mining Revenues for the Public Good—Ensuring Transparency and Distribution Equity. *Marine Policy* 95 (September): 332–336. https://doi.org/10.1016/j.marpol.2017.02.010.

Seas at Risk. 2018a. Environmental Groups Call for a U-Turn on Deep Sea Mining.

———. 2018b. Public Consultation on First Deep Sea Mining Equipment Test in the Clarion Clipperton Zone. September 2018.

Smits, Coco C.A., Judith van Leeuwen, and Jan P.M. van Tatenhove. 2017. Oil and Gas Development in Greenland: A Social License to Operate, Trust and Legitimacy in Environmental Governance. *Resources Policy* 53: 109–116. https://doi.org/10.1016/j.resourpol.2017.06.004.

Toonen, Hilde M., and Jan P.M. van Tatenhove. 2020. Uncharted Territories in Tropical Seas? Marine Scaping and the Interplay of Reflexivity and Information. *Maritime Studies* 19 (3): 359–374. https://doi.org/10.1007/s40152-020-00177-z.

Voyer, Michelle, and Judith van Leeuwen. 2019. 'Social License to Operate' in the Blue Economy. *Resources Policy* 62. https://doi.org/10.1016/j.resourpol.2019.02.020.

Warner, Robin. 2020. International Environmental Law Principles Relevant to Exploitation Activity in the Area. *Marine Policy* 114 (April). https://doi.org/10.1016/j.marpol.2019.04.007.

Willaert, Klaas. 2020a. Effective Protection of the Marine Environment and Equitable Benefit Sharing in the Area: Empty Promises or Feasible Goals? *Ocean Development & International Law* 51 (2): 175–192. https://doi.org/10.1080/00908320.2020.1737444.

————. 2020b. On the Legitimacy of National Interests of Sponsoring States: A Deep Sea Mining Conundrum. *The International Journal of Marine and Coastal Law* October: 1–18. https://doi.org/10.1163/15718085-BJA10011.

Zalik, Anna. 2018a. Mining the Seabed, Enclosing the Area: Ocean Grabbing, Proprietary Knowledge and the Geopolitics of the Extractive Frontier beyond National Jurisdiction. *International Social Science Journal* 68 (229–230): 343–359. https://doi.org/10.1111/issj.12159.

————. 2018b. Mining the Seabed, Enclosing the Area: Proprietary Knowledge and the Geopolitics of the Extractive Frontier beyond National Jurisdiction. *International Social Science Journal* 15. https://doi.org/10.1111/issj.12159.

CHAPTER 5

Transboundary Regionalization at European Seas

Abstract This chapter analyses the liquid institutionalization of transboundary maritime regionalization at the scale of European seas. The chapter starts with defining transboundary regionalization and the differences between transboundary and cross-border cooperation. To answer the question what types of liquid institutionalization can be distinguished, the chapter analyses the emerging transboundary regionalization governance arrangements, of Transboundary Maritime Spatial Planning (TMSP) in the North Sea, the Baltic Sea and the Atlantic Ocean, Sea Basin Strategies and Marco-Regional Strategies (for the Baltic Sea Region (2009) and for the Adriatic and Ionian Region (2014)). The analysis of transboundary regionalization governance arrangements showed two very different faces, and types of liquid institutionalization. First, analysis identified the informal governance experiments of Transboundary Maritime Spatial Planning projects and, secondly, the EU initiated formal governance processes of the Macro Regional Strategies.

Keywords Transboundary regionalization • Marine/Maritime Spatial Planning • Sea Basin Strategies • Macro Regional Strategies • Transboundary Maritime Spatial Planning • Cross-border cooperation • Transboundary cooperation • EUSBSR • EUSAIR

© The Author(s), under exclusive license to Springer Nature Switzerland AG 2022
J. P. M. van Tatenhove, *Liquid Institutionalization at Sea*,
https://doi.org/10.1007/978-3-031-09771-3_5

5.1   Introduction

This chapter analyses the institutionalization of different processes of transboundary regionalization at European seas. Section 5.2 defines transboundary regionalization based on a discussion of the concepts of regionalization, and cross-border and transboundary cooperation. Section 5.3 describes and analyses two forms of transboundary regionalization; Transboundary Maritime Spatial Planning (TMSP), and EU Regional Strategies (Sea-Basin Strategies (SBS) and Macro-Regional Strategies (MRS)), and the emerging governance arrangements in these forms of transboundary regionalization, followed in Sect. 5.4 with an analysis of the specific forms of liquid institutionalization of each of these examples of transboundary regionalization.

5.2   How to Understand
Transboundary Regionalization?

Regionalization can be understood in different ways. Eliasen et al. (2015) understand regionalization as the decentralization of fisheries management from the EU to the regional sea level, with an increased involvement of stakeholders. Maier and Markus (2013: 72) distinguish between different forms of regionalization, ranging from region-specific measures to "more institutionalized, region specific coordination and articulation of interests and cooperative implementation." In 1979 the EU Court of Justice set the frame in which regionalization could develop. Regional management can be organized at the EU level or the Member State level. "Any type of inclusion of non-EU or non-Member States actors in the legislative process would—due to the Commission's ultimate right of initiative—be restricted to giving input in the pre-decision phase of the process" (Maier and Markus 2013: 72). With the regionalization of international law, Regional Seas Programmes have been developed, with regional institutions providing the link between the "global and national or local level of governance so that there is a system of co-responsibility according to the principle of subsidiarity" (Rochette et al. 2014; 109).

Another form of regionalization is macro-regionalization, defined by Gänzle and Kern (2016: 6) "as processes, eventually underwritten by macro-regional strategies and underpinned by a single strategic approach. This approach must aim at the construction of functional and transnational spaces among those (administrative) regions and municipalities at the subnational level of EU member and partner countries that share a sufficient number of issues in common." They see macro-regionalization as a shift from territorial towards functional regions, providing political opportunities for sub-national authorities and civil society. All these definitions recognize a regional level, in this case the regional sea level, at which specific formal and informal interactions and negotiations take place, and where UN institutions, EU institutions, Member States and sub-national authorities have specific responsibilities in processes of policy-making and governance. Van Tatenhove and Van Leeuwen (2015: 191–192) also distinguish other forms of regionalization, rather than governmental cooperation alone. Processes of marine governance at the regional sea level concern the geographical aspects of the ecosystem and the maritime activities across multiple boundaries. They distinguish three forms of regionalization of marine governance. First, regionalization as cooperation of governmental actors at the level of the regional sea. This form refers to the need for neighbouring countries and governmental actors on the supra and international level to cooperate at the regional seas level in order to coordinate and implement policies. Second, regionalization as the empowerment of non-governmental actors at the regional seas level. This form of regionalization is about creating institutions on the regional sea level, to accommodate for example stakeholder involvement, such as the Advisory Councils in the Common Fisheries Policy (CFP). Third, regionalization as an organizing principle of maritime activities at the regional seas level. This form is about organizing maritime activities at the regional sea level, such as the development of an Offshore Renewable Power Grid, combining of surveillance activities of coast guards or the development of a network of Marine Protected Areas. Building on these three forms of regionalization Soma et al. (2015) define regionalization as "processes of cooperation and integration in which territorial spaces are (re)defined and political spaces are contested and (re)composed at the regional level." In other words, regionalization can be defined on interaction (agency) and structural levels. On the level of interactions, regionalization is the result and outcome of negotiations between (interdependent) public and private actors about the ordering of spatial processes, and the organization of maritime activities within certain spaces and regions. On the structural level, regionalization refers to the specific institutionalization of governance arrangements needed to accompany these social and political activities of spatial ordering. Territorial spaces and boundaries are defined in social and political interactions. In this sense, regionalization refers to a power process in which regions/areas—as political spaces—are (re)composed and contested, based on an unequal division of resources. More specific the power process to compose and to create regions as political spaces focuses on the capacity of actors to mobilize and use resources, to cooperate on a regional level, the empowerment of non-governmental actors in the design and implementation of regionalization processes and the ability to designate and organize maritime activities spatially.

"**Transboundary**" is the second core concept of this chapter. Transboundary is a crucial concept because planning and political processes at the regional sea level are transboundary in nature, for example crossing and transcending the borders of individual nation states. Li and Jay (2020) define transboundary as the "engagement of multiple entitles (e.g. countries, states, provinces) across one ecosystem, who also do not necessarily share a common border" (pp.3). This definition raises the question what the differences are between transboundary cooperation and **cross-border** cooperation. In a special issue in European Planning Studies Nienaber and Wille (2020) bring together a collection of papers with examples of cross-border cooperation (mainly on land). According to the authors, cross-border cooperation requires new ways of spatial planning, which are not set according to national borders and can be seen as a measure to overcome the national setting, involving national, local, and regional planning institutions of both sites of the border negotiating the planning process. The chosen relational perspective of cross-border focuses on informal cross-border development (instead of formal plans) and combines network, governance and territorialization.[1] The main difference between cross-border and transboundary cooperation is that cross-border cooperation is collaboration between two neighbouring countries, while transboundary cooperation is collaboration between public and private actors for example at the level of the regional sea, not necessarily dependent on them being neighbours of each other. I use the term "transboundary" to encompass planning, governance and power processes at the regional sea level in which non-neighbouring riparian states also are involved. In other words, transboundary regionalization is a form of regionalization at the level of the regional sea in which governmental and

---

[1] This is based on (Healey 2006). The multiple networks existing in a cross-border region lead to different and also new forms of cross-border governance with different power relations among the network actors, that then result in "institutional sites, with particular material geographies" and "relational complexity."

non-governmental actors, organized at different levels (global, EU, national, sub-national), negotiate about the ordering of territorial spaces, the legitimacy of maritime activities, their spatial positioning, and the governance arrangements needed to govern these processes of ordering and positioning.

5.3   DIFFERENT FORMS OF TRANSBOUNDARY REGIONALIZATION IN EUROPE

This section discusses the institutionalization of two forms of transboundary regionalization at seas: Transboundary Maritime Spatial Planning (TMSP) and EU Regional Strategies (Sea-Basin Strategies and Macroregional Strategies for the Baltic Sea and the Adriatic and Ionian Sea). These cases have been selected because the governance and planning processes take place at the level of the regional sea, while the planning and decision-making processes are the result of negotiations and collaboration between governmental and non-governmental actors at different levels.

For each of these forms of regionalization, the emerging transboundary governance arrangements will be described and analysed. The central aim of this chapter is to explore the different types of liquid institutionalization these forms of transboundary regionalization show.

5.3.1   Transboundary Maritime Spatial Planning

The Development of MSP
Over the last 15 years, Marine/Maritime Spatial Planning (MSP) has become the dominant management discourse to deal with conflicting human uses at sea and to endorse sustainable development at sea. Several publications have presented and propagated MSP as a management tool: the rational organization of the use of marine space, balancing these demands and to achieve ecological, economic and social objectives in a planning process and to be specified through a political process (Douvere 2008; Douvere and Ehler 2009; Douvere and Ehler 2007). The 2008 Special Issue in Marine Policy "The role of Marine Spatial Planning in implementing Ecosystem-based, Sea Use Management" edited by Fanny Douvere and Charles Ehler was one of the first scientific comprehensive attempts to give insight into a range of aspects of MSP. These include drawing attention to the relationship of MSP with Ecosystem Based

Management (EBM), the missing layer of MSP (St. St Martin and Hall-Arber 2008), the international legal framework of MSP (Maes 2008), the engagement of stakeholders in MSP (Pomeroy and Douvere 2008) and MSP at the high seas (Ardron et al. 2008). From the beginning of the 2000s, governments, the EU and the UN embraced the idea of MSP and started several initiatives. In 2007, the EC proposed in its Blue Paper[2] an Integrated Maritime Policy for the European Union, "based on the clear recognition that all matters relating to Europe's oceans and seas are inter-linked, and that sea-related policies must be developed in a joined-up way if we are to reap the desired results" (EC 2007: 2). The main legitimation for the EC was that a more collaborative and integrated approach was needed to deal with the "increasing competition for marine space and the cumulative impact of human activities on marine ecosystems" (EC 2007: 4) and to overcome the inefficiencies, incoherencies and conflicts of use caused by fragmented decision-making in maritime affairs. The development of MSP in the EU started with defining it as a policy tool in its "Roadmap for Maritime Spatial Planning: Achieving Common Principles in the EU" allowing "public authorities and stakeholders to coordinate their actions and optimize the use of marine space to benefit economic development and the marine environment" (EC 2008: 2) to a Maritime Spatial Planning Directive in 2014 (Directive 2014/89/EU). In 2009, Ehler and Douvere developed a step-by-step approach for MSP towards Ecosystem Based Management approach (UNESCO 2009). This publication presented a straightforward guide for professionals responsible for the planning and management of marine areas and their resources, how to set up and apply MSP in ten steps.

In policy practices and scientific debates, there is a broad consensus about a rationalist conception of MSP. In this framing, MSP is a tool or instrument to allocate and distribute marine space to different (competing) uses and to manage the marine environment in a sustainable way. This is reflected in definitions of the "United Nations Educational, Scientific and Cultural Organization" (UNESCO) (UNESCO 2009) and the

[2] With regard to the marine environment, the European Commission's Strategic Objectives for 2005–2009 state that "in view of the environmental and economic value of the oceans and the seas, there is a particular need for an all-embracing maritime policy aimed at developing a thriving maritime economy and the full potential of sea-based activity in an environmentally sustainable manner" (COM (2005) 12 final). This commitment resulted in the Green paper (EC 2006) in June 2006 and after a consultation round of one year with stakeholders in the Blue paper (EC 2007).

European Union (EC 2008), which emphasize MSP as a tool for improved decision-making (EU), to arbitrate between (conflicting) sectoral interests in a political process (UNESCO), and to achieve sustainable use of marine resources.

This dominant MSP discourse is built upon four different storylines (Vince et al. 2013; Douvere and Ehler 2009; Soininen et al. 2015; Flannery et al. 2016, 2020; van Tatenhove 2017). First, "MSP as a (neutral) referee" to solve conflicts between different sectoral maritime interests. Second, "MSP enables sectoral integration" and while incorporating policies from different layers of government, offering opportunities for a more strategic and forward-looking framework for all uses at sea. Third, "MSP is sustainable." From the integrative storyline, MSP anticipates and addresses future resource demands in a sustainable manner and is able to manage the threats of overexploitation in territorial seas of individual states. Soininen et al. (2015) emphasize different forms of sustainability. On the one hand, the economic and social responsibility related issues of sustainability linked to the emerging concept of the "Blue Economy" and coastal communities, and, on the other hand, as an instrument of environmental protection, emphasizing environmental sustainability within the goal of Ecosystem-Based Management (EBM). Fourth, MSP framed "as a way to overcome inefficiencies." These inefficiencies arise from fragmented governance structures and regimes. The dominant MSP discourse presents maritime spatial planning processes as systematic and rational, and positions MSP as a panacea to deal with conflicts at sea in a sustainable way, to overcome the "inefficiencies" that arise from sector-based management and fragmented governance regimes, presenting an integrated solution for the inefficiencies, and increasing the legitimacy of marine policies by involving stakeholders.

However, recent studies show that the dominant storylines of MSP are no panaceas. According to Ellis and Flannery (in Flannery et al. 2016) distributional impacts of MSP are insufficiently taken into consideration. Because power is not acknowledged in MSP processes one could assume that MSP processes reflect existing power structures and that powerful interests (such as maritime industries or INGOs) shape the outcomes and planning processes, also affecting the inclusion and exclusion of stakeholders. St Martin and Hall-Arber (2008) show that the social landscape of the marine environment is undocumented and remains a "missing layer" in MSP decision-making.

MSP has developed rapidly over the last decade. Sixty countries have MSP processes underway and expect to have this completed between 2020 and 2025. Four countries have completed their plan but that has not been implemented yet (Belize, Canada (Beaufort Sea, Atlantic), Mexico and Portugal), five countries have completed and implemented their plan (Australia, Canada (Pacific), China, Germany, USA (MA, OR, RI). Only three countries have had their first (or second) five-year review of an implemented plan (Belgium, the Netherlands and Norway) (Smith 2018).

The Development of an Experimental TMSP Governance Arrangement
According to the Maritime Spatial Planning Directive (MSPD; Directive 2014/89/EU) EU Member States (MSs) have to establish a maritime spatial plan 31 March 2021 the latest (art 15(3). Although the MSs are responsible for developing their own maritime spatial plan and setting up their own maritime planning processes, the MSPD emphasizes the need for cooperation. "Member States should consult and coordinate their plans with the relevant Member States and should cooperate with third-country authorities in the marine region concerned in conformity with the rights and obligations of those Member States and of the third countries concerned under Union and international law" (preamble (20)). One of the minimum requirements for maritime spatial planning is to "ensure trans-boundary cooperation between Member States in accordance with Article 11" (art 6(f)). According to Article 11 "Member States bordering marine waters shall cooperate with the aim of ensuring that maritime spatial plans are coherent and coordinated across the marine region concerned. Such cooperation shall take into account, in particular, issues of a transnational nature," including cooperation with third countries (article 12).

In earlier research (van Tatenhove 2017: 785), I argued the necessity of MSP as transboundary planning and decision-making processes. Main reasons for this argument are firstly that the natural environment is fluid, and that marine ecosystems move across different administrative-political borders. Secondly, many marine resources and maritime activities are cross-border and mobile in nature. Effective planning and management require a collaborative approach for neighbouring and non-neighbouring jurisdictions. Thirdly, physical boundaries are generally absent in this remote, dynamic, and graded environment, making it difficult to contain many activities and their impacts within administrative territories, and finally, MSP is generally conducted at larger geographical scales, including considerations of regional and land–sea interaction. Transboundary thinking is therefore becoming part of the rationale of MSP and an expression of the distinctive nature of spaces and regions to be planned. However, the development, planning and implementation of Transboundary Maritime Spatial Planning (TMSP) take place in a fragmented institutional setting, characterized by high ambiguity, conflicting regime complexes and emerging network states (van Tatenhove 2016).

Despite the recognition of the importance to ensure transboundary cooperation in the MSPD, existing institutional barriers (such as institutional inertia, path dependency due to specific national jurisdictions and governmental institutional systems, procedural obstacles, and different national planning systems) hamper the development of TMSP practices. TMSP is not yet an existing governmental planning practice; it is in an experimental phase to explore the constraining conditions of TMSP and to develop best TMSP practices. This is not yet anchored in formal procedures and regulations in countries or in regional sea governance institutions. In the last decade, many projects (co-)financed by DG MARE have been developed to explore the enabling and constraining conditions of TMSP, more specifically cross-border and transboundary cooperation and planning. The governance architecture of these projects is developed by researchers, in cooperation with national authorities. Examples of these projects are *MASPNOSE—"Maritime Spatial Planning in the North Sea"* (Pastoors et al. 2012; Hommes 2012), *Plan Bothnia* (Hermanni Backer and Frias 2013; H. Backer et al. 2013), *BaltSeaPlan* (Schultz-Zehden et al. 2013; Gee et al. 2013; Käppeler et al. 2012), *Baltic SCOPE* (Kull et al. 2017; Moodie et al. 2021; Kull et al. 2021), *Pan Baltic SCOPE* (Cedergren et al. 2019; Moodie et al. 2021; Kull et al. 2021), *Study on International Best Practices for Cross-Border MSP* (Cross-border MSP Study) (Kull et al. 2021), *TPEA—Transboundary Planning in the European Atlantic* (Jay et al. 2016; Almodovar et al. 2014), *SIMNORAT— Supporting Implementation of Maritime Spatial in the Northern European Atlantic Region* (Gómez-Ballesteros et al. 2021) and *ADRIPLAN— Developing a Maritime Spatial Plan for the Adriatic and Ionian Region* (http://adriplan.eu/, visited 15/10/2021).

These projects show an emerging experimental TMSP governance arrangement. The main characteristics of this governance arrangement are that it does not reflect formal maritime spatial planning processes and the development of a joint planning practice, rather that these projects are informal and experimental governance practices to explore the possibilities of developing institutional TMSP governance arrangements within the given institutional setting of the EU (implementation of the MSPD). Governmental and non-governmental actors in the North Sea, Baltic Sea, Mediterranean and European Atlantic have an interest in developing transboundary initiatives but perceive the development of joint formal plans not as realistic, due to the existing institutional (national) barriers and sovereignty issues hampering transboundary cooperation and collaboration (van Tatenhove 2017). This is reflected in the TMSP projects. For example the mission of SIMNORAT is not to develop a plan, but to develop and test aspects of MSP processes to produce guidelines and recommendations for MSP processes in the involved countries. Also Plan Bothnia did not aim at developing a politically accepted plan, but to explore the possibilities of long-term transboundary cooperation between Sweden and Finland.

The dominant discourse of this experimental TMSP governance arrangement, as reflected in the projects mentioned, is "learning by doing and experimentation in a controlled setting." The aim of all these projects is to explore the opportunities and constraints of transboundary cooperation and TMSP (MASPNOSE, BaltSeaPlan, Baltic SCOPE, SIMNORAT, ADRIPLAN), to try out strategic and realistic transboundary spatial planning approaches (Plan Bothnia, TPEA), to explore Land-Sea Interactions (LSI) and to develop connectivity thinking (Pan Baltic Scope, BaltSeaPlan), to develop strategies of stakeholder involvement (MASPNOSE, Baltic SCOPE, BaltSeaPlan), and improve the cooperation between countries (SIMNORAT).

Each of these experimental projects formulated lessons learned and best practices for TMSP, transboundary/cross-border cooperation and the involvement of stakeholders. The impressive number of innovative ideas in these experimental planning "laboratories" sets the scene for future institutional rules about transboundary cooperation, collaboration between governmental and non-governmental actors and the institutions needed for that. Researchers are core actors in these experimental projects. They perform the role of programmers by setting-up the projects, by formulating the objectives, and selecting which governmental and non-governmental actors should be included. As switchers, they bring different national MSP arrangements together and formulate lessons learned and best practices based on the outcomes of existing projects.

5.3.2   EU Regional Strategies

Introduction

The EU has developed two forms of regional strategies: a Sea Basin Strategy (SBS) and Macro-Regional Strategy (MRS). **A Sea-basin strategy** is a "structured framework of cooperation in relation to a given geographical area, developed by Union institutions, Member States, their regions and where appropriate third countries sharing a sea basin; a sea basin strategy takes into account the geographic, climatic, economic and political specificities of the sea basin" (Common Provisions Regulation (EU) No 1303/2013). In other words, a sea-basin strategy is a maritime strategy for a regional sea. It is marine- and maritime-centred with the objective to provide a more coherent approach to maritime issues, with increased coordination between different policy areas (Interact 2014). Since 2007, The Commission has developed the following SBSs (see Fig. 5.1): Black Sea Synergy (2007),[3] first Communication on the Arctic Region (2008), North Sea Forum (2009), Communication to Improve Maritime Governance in the Mediterranean (2009), the Atlantic Strategy (2011), Maritime Strategy for the Adriatic and Ionian Seas (2012). Characteristic for each of these SBSs is that they formulate the possibilities for an integrated maritime strategy involving different stakeholders.

In 2012, the EC published a Maritime Strategy for the Adriatic and Ionian seas (EC, COM(2012) 713 final). This SBS sets out a framework to move towards a coherent maritime strategy and corresponding Action Plan by 2013. This communication is also seen as a first component in the preparation of MRS for the Adriatic and Ionian region and provides a framework to adapt the Integrated Maritime Policy to the needs and potential of the natural resources and socio-economic fabric of the Adriatic Ionian marine and coastal areas (pp. 3).

A **Macro-Regional Strategy** (MRS) is "an integrated framework endorsed by the European Council, (…) to address common challenges faced by a defined geographical area relating to Member States and

---

[3] The Black Sea Synergy "encourages a bottom-up approach to project development, identifying and supporting the needs, priorities and aims of partners in the region, and what they want to do together. The key elements of the Black Sea Synergy include building confidence, fostering regional dialogue and achieving tangible results for states and citizens in the region. A stable, secure, resilient and prosperous Black Sea region is in the direct interest of its citizens and of the EU overall" (https://eeas.europa.eu/diplomatic-network/black-sea-synergy/346/black-sea-synergy_en, visited 7 June 2021) (EU External Action Service, 2021).

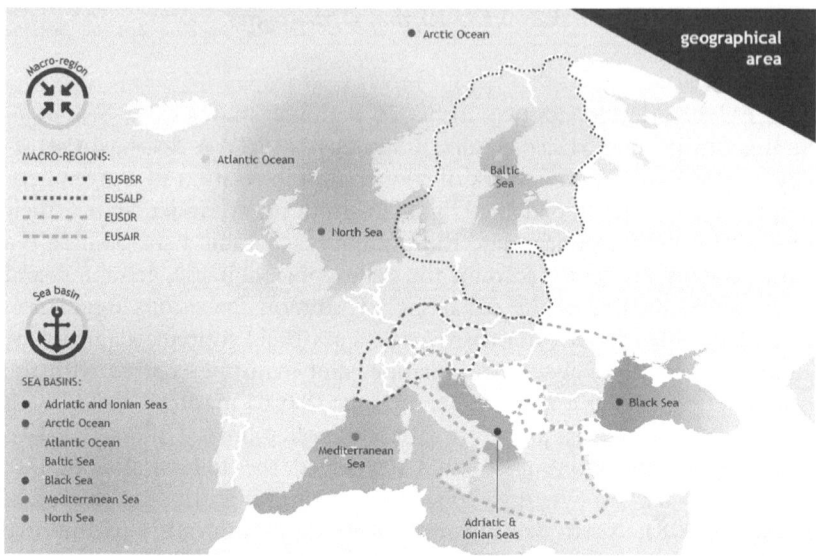

Fig. 5.1 EU Sea-Basin Strategies and Marco-Regional Strategies (Interact September 2014)

countries located in the same geographical area which thereby benefit from strengthened cooperation contributing to achievement of economic, social and territorial cohesion" (Common Provisions Regulation (EU) No 1303/2013). Since 2009, the EU adopted four MRSs (see Fig. 5.1): The European Union Strategy for the Baltic Sea Region (EUSBSR)[4]; the European Strategy for the Adriatic-Ionian Region (EUSAIR) (2014),[5] the European Strategy for the Danube region (EUSDR) (2011)[6] and the European Strategy for the Alpine Region (EUSALP) (2015).[7]

After discussing the similarities and differences of the SBSs and MRSs in general (Sect. 5.3.2.2), Sect. 5.3.2.3 describes and analyses the two sea-related MRSs (EUSBSR and EUSAIR). The reason for this selection is that both MRSs present forms of transboundary regionalization as defined in this chapter, that planning processes are based on negotiations between

[4] http://balticsea-region.eu/ visited 18 June 2021.
[5] https://www.adriatic-ionian.eu/ visited 18 June 2021.
[6] https://danube-region.eu/ visited 18 June 2021.
[7] https://www.alpine-region.eu/eusalp-eu-strategy-alpine-region visited 18 June 2021.

public and private actors and that in these (collaborative) interactions problems are defined and possible solutions are formulated. These strategies have the opportunity to develop new (institutional) governance arrangements and to negotiate future scenarios, based on land–sea interactions.

Similarities and Differences Between Sea Basin Strategies and Macro-Regional Strategies
Both SBS and MRS focus on common issues, solutions, and actions of strategic relevance for EU Member States and non-EU countries in a regional sea region. Both strategies encourage strategic cooperation and coordination among policies, institutions and funding sources, while implementation requires an integrated approach establishing cross-sectoral cooperation and coordination mechanisms as well as multi-stakeholder dialogue (INTERACT 2014). Besides these similarities, the SBS and MRS differ on the following aspects. Firstly, a SBS is oriented to the regional sea and its maritime activities and seeks to provide a more coherent approach to maritime issues, with increased coordination between different policy areas. A MRS takes also into consideration land-sea connections and the land dimension of sea-based activities (such as port development and the relation with shipping, fisheries and tourism, and hinterland connection of ports and land-based pollution by agriculture and industries) of a regional sea region and aims to achieve economic, social and territorial cohesion. Secondly, both strategies should contribute towards achieving Europe 2020 Strategy targets with a specific focus on Blue Growth for the SBS. This is reflected in the fact that for the SBS the Directorate General for Maritime Affairs and Fisheries (DG MARE) is responsible while for the MRS, the Directorate General for Regional and Urban Policy (DG REGIO) is responsible.

Macro-Regional Strategies: The Examples of EUSBSR and EUSAIR
With MRSs the EU has started to develop its own "regional institution" at a new "intermediate territorial level" between the EU and MSs (Gänzle and Mirtl 2019). EU MRS is a form of functional cooperation in a territorial context, which is subject to the principle of "Three Nos"; no additional costs; no new institutions and no specific legislation for the macro-region. Scholars define MSR as a form of experimentalist governance (Gänzle and Mirtl 2019) and as "soft-spaces" of governance (Faludi

2010) with the characteristics of informality, variability and flexibility of the institutional structures, different thematic and stakeholders.

The European Union Strategy for the Baltic Sea Region (EUSBSR)

The EUSBSR was the first MSR in Europe, and is a unique platform for cooperation and coordination between eight EU Member States (Denmark, Estonia, Finland, Germany, Latvia, Lithuania, Poland, Sweden) and four non-EU countries (Belarus, Iceland, Norway, Russia) in the Baltic Sea region (see Fig. 5.2).

The aim of the EUSBSR is to develop an integrated framework to address common challenges, based on functional cooperation in the Baltic Sea region based on the "Three Nos" principle.

After revising and re-ratification of the Helsinki Convention in 1992, major implementation gaps were observed (Gløersen et al. 2019), such as sea pollution from terrestrial sources (agriculture, industry, municipal waste water, etc.), the need to develop integrated programmes and policies with neighbouring countries. Environmental challenges were an important reason for the European Council (December 2007) to invite the EC to present an EU strategy for the Baltic, and the Swedish presidency of the Council (second half of 2009) actively promoted the Strategy. Important reasons were to deal with environmental issues and to improve regional cooperation.[8] The first EUSBSR Action Plan (June 2009) foresaw environmental measures at the level of the Baltic Sea Region in coordination with pre-existing transnational initiatives, such as the Baltic Sea Action Plan (BSAP) of HELCOM (2007), and the Northern Dimension Environmental Partnership (NDEP from 2001) (pp. 62). An important rationale to develop the EUSBSR is that the implementation mechanisms in the BSAP are perceived as insufficient, while compared to the NDEP, the EUSBSR promotes a more integrated perspective on an "environmental neighbourhood policy." The EUSBSR has three main objectives: Save the Sea; Connect the Region; and Increase Prosperity.

In the EUSBSR governance structure EU MSs, non-EU states and Baltic Sea institutions (such as the Council of the Baltic Sea States (CBSS), Baltic Marine Environment Protection Commission—Helsinki

---

[8] Organizations such as HELCOM, the Council of Baltic Sea States (CBSS), Baltic 21 (regional Agenda 21 for the Baltic Sea region) struggled to find effective forms of regional cooperation.

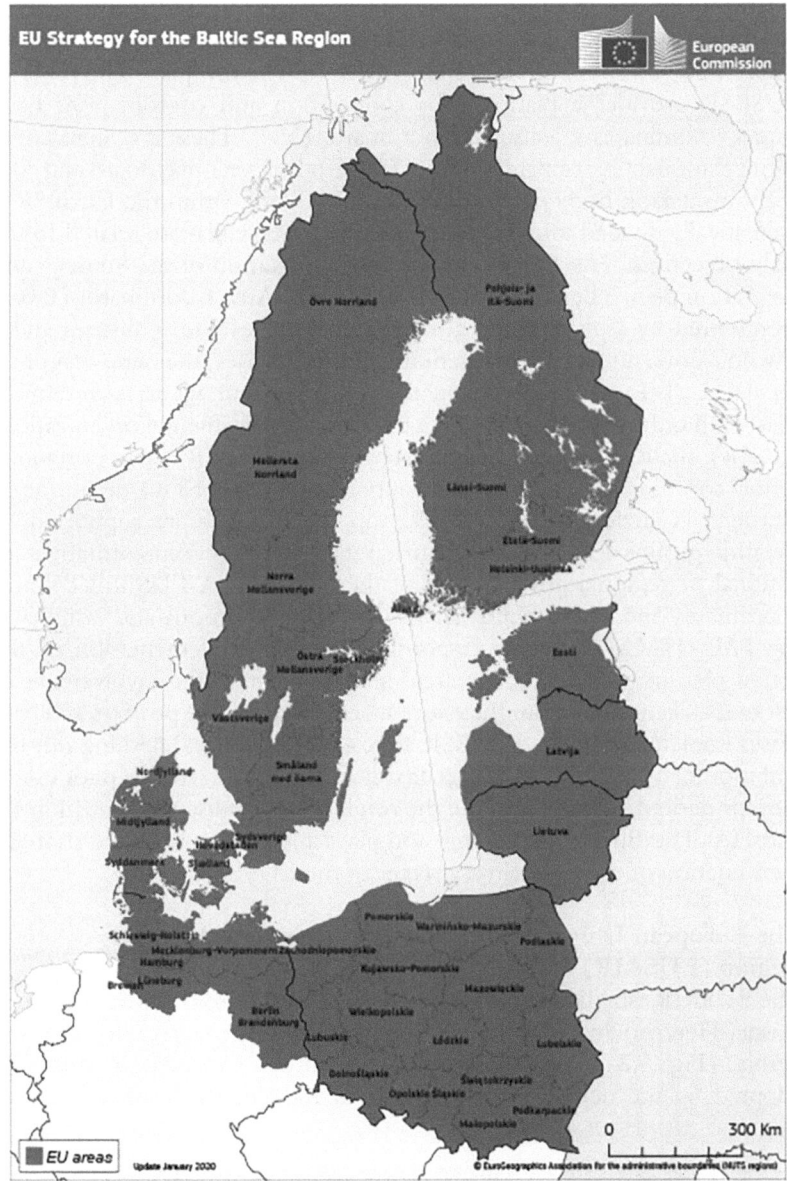

Fig. 5.2   The EUSBSR area. (Source: https://ec.europa.eu/regional_policy/en/ policy/cooperation/macro-regional-strategies/baltic-sea/, visited 16/03/2021)

Commission (HELCOM), Vision and Strategies around the Baltic Sea (VASAB)) play a key role in developing and implementing the Strategy (EC 2020b) (EC 2020c) (EU/EUSBSR, Action plan 2021). The EUSBSR provides a platform for cooperation and coordination. Each country nominates a National Coordinator (NC). These are represented in the National Coordinators Group (NCG) the Executive Board and core decision-making body of the EUSBSR. The work within the EUSBSR is thematically divided into 14 Policy Areas (PAs),[9] which are related to the main objectives. The PAs ensure the implementation of the Strategy and the Action plan. They are managed by a Policy Area Coordinator (PAC), represented by (sub)national government agencies and ministries and a few non-governmental organizations. "In most cases, thematic coordinators (PAC, JvT) represent (sub)national government agencies and ministries, with only a few appointments from non-governmental organizations (NGOs); the latter applying exclusively to the EUSBSR where horizontal action coordinators (HACs) were appointed by the NGO organization Norden" (Gänzle and Mirtl 2019) (pp. 246). PACs are supported by steering groups (SG); steering group members are representatives of national or regional governments of the EUSBSR. A EUSBSR country coordinates one of the policy areas or horizontal actions and nominates the PAC/HAC. These are responsible for the implementation of the action plan in their respective areas and to facilitate the involvement of relevant stakeholders from the entire macro-region. The projects and processes implementing the EUSBSR are called "flagships," fleshing out the ambition of a PA or HA in a specified field. Flagships serve as pilot examples for desired change; they are the result of policy discussions within PA and HA. The European Commission plays a leading role in the strategic coordination of the key delivery stages of the EUSBSR.

The European Union Strategy for the Adriatic and Ionian Region (EUSAIR)

The EUSAIR established a level-playing field with third countries (Albania, Bosnia-Herzegovina, Croatia, Greece, Italy, Montenegro, Slovenia and Serbia) (Fig. 5.3). The initiative draws from the "Ancona Declaration'" adopted by the decision-making body of the Adriatic Ionian Initiative

---

[9] The 14 policy areas are: PA Nutri, PA Hazards, PA Bio-economy, PA Ship, PA Safe, PA Transport, PA Energy, PA Spatial Planning, PA Secure, PA Tourism, PA Culture, PA Innovation, PA Health and PA Education.

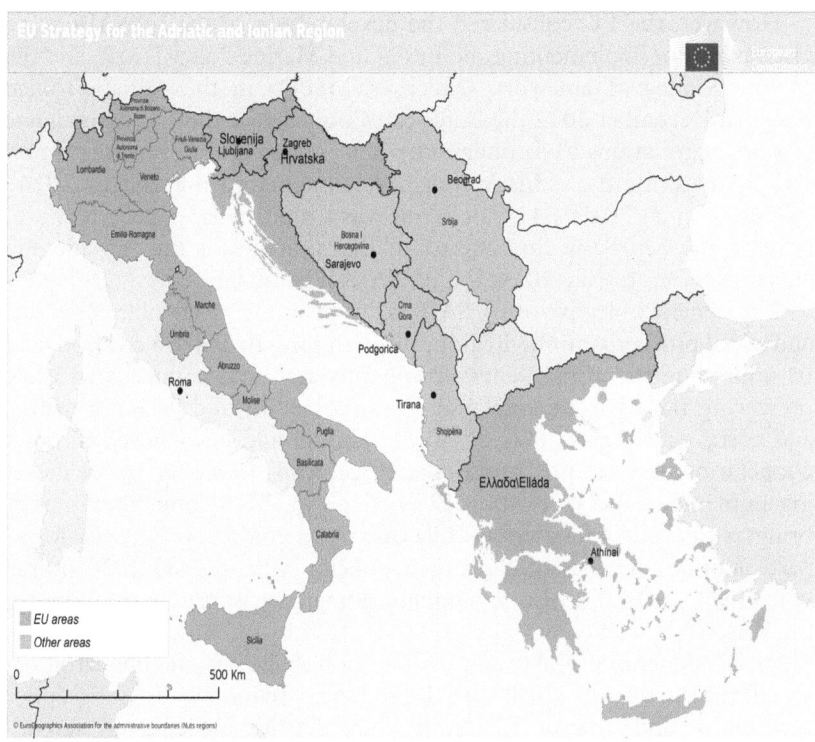

Fig. 5.3 Map of the EUSAIR. (Source EU, COM (2014) 357 final)

(AII[10]): the Adriatic-Ionian Council (AIC). The AII signed the Ancona declaration "which seeks to strengthen regional cooperation to promote political and economic" (Gänzle and Mirtl 2019) (pp. 244). The AII is an inter-governmental platform of cooperation that encompasses environmental protection but extends also to other fields, such as economic development, land transport connections, health and cultural cooperation, tourism development and maritime cooperation (Gløersen et al. 2019: 63).

In the Ancona Declaration, the Adriatic-Ionian MRS is considered as a way to enhance regional cooperation in different fields.

---

[10] The AII is established May 2000 on the Summit on Development and Security on the Adriatic and Ionian Seas.

However, the EC considered the development of the EUSAIR as an instrument for implementing the Integrated Marine Policy (IMP) and the Marine Strategy Framework Directive (MSFD) in the Adriatic-Ionian basin. In December 2012 the European Council gave the EC the mandate to start negotiations to formulate the Action Plan and to present "a new EU Strategy for the Adriatic-Ionian Region before the end of 2014" (Gløersen et al. 2019: 64). The result was a multi-pillar approach and to develop the Sea Basin Strategy to an EU strategy for the Adriatic and Ionian Region. In July 2014, the EUSAIR was established.

The general objective of EUSAIR is "to promote sustainable economic and social prosperity in the Region, through growth and job creation, and by improving the attractiveness, competitiveness and connectivity, while preserving the environmental and ensuring healthy and balanced marine and costal ecosystems" (EC 2014: 3). This should be achieved through cooperation between the countries, and according to the EC by "reinforcing implementation of existing EU policies in the Region, the Strategy brings a clear EU added value while offering a golden opportunity for all participating countries to align their policies with the EU-2020 overall vision" (EC 2014: 3). Also, a geopolitical objective is formulated to bringing Western Balkan countries closer to the EU by offering them opportunities for cooperation and to address common challenges and opportunities specific to the Region. Challenges are the socio-economic disparities, poor accessibility and (transport) infrastructure deficits, inadequate interconnection of energy grids, environmental threats, risks related to climate change and administrative and institutional issues. Opportunities are the potential for smart, sustainable, and inclusive growth, such as blue growth initiatives, improving the land-sea connectivity, preserving the cultural and natural heritage and biodiversity, sustainable development of tourism.

According to Gløersen et al. (2019: 64), the EUSAIR is shaped by two important factors. Firstly, the efforts of Italy to develop the MRS as an integrated development strategy with a marine/maritime dimension. Secondly, DG REGIO (instead of DG MARE) took the lead of the process, resulting in a strategy emphasizing both maritime/marine issues and fields as transport and tourism. The earlier Maritime Strategy (sea-basin strategy) is reflected in the pillars "blue growth" and to "environmental quality."

EUSAIR is structured around the following pillars: "Blue Growth" (coordinated by Greece and Montenegro); "Connecting the Region": transport and energy networks (coordinated by Italy and Serbia);

"Environmental quality" (coordinated by Slovenia and Bosnia-Herzegovina), and "Sustainable tourism" (coordinated by Croatia and Albania). "Capacity building" and "Research and Innovation" are defined as crosscutting aspects. For each pillar, the following main objectives are formulated:

- *Blue Growth*: to drive innovative maritime and marine growth in the Region by promoting sustainable economic growth and jobs as well as business opportunities in the Blue economy sectors (EC 2020a: 6). The focus is on "blue technologies," "fisheries and aquaculture" and "maritime and marine governance and services."[11]
- *Connecting the Region*: "to improve connectivity within the Region and with the rest of Europe in terms of transport and energy networks. This requires thorough coordination of infrastructure works and improved operation of transport and energy systems between the countries in the Region" (EC 2020a: 22). The focus is on "maritime transport," "intermodal connections to the hinterland" and "energy networks." Since the core transport and energy axes are macro-regional and even European in the perspectives of the Trans-European Networks for transport (TEN-T) and energy (TEN-E), while the national railway and energy markets are fragmented and too small to attract investors, regional coordination is crucial to overcome bottlenecks and to promote connectivity.
- *Environmental Quality*: "to address the issue of environmental quality, with respect to marine, coastal and terrestrial ecosystems in the Region" (EC 2020: 37). Environmental quality is essential for underpinning human activities in the Region and for ensuring economic and social well-being. The focus is on "the marine environment" and "transnational terrestrial habitats and biodiversity."
- *Sustainable Tourism*: to develop "the sustainable and responsible tourism potential of the Adriatic-Ionian Region, through innovative and quality tourism products and services. It also aims at promoting responsible tourism behaviour on the part of all stakeholders (wider public, local, regional and national private and public actors, tourists/visitors) across the Region" (EC 2020: 52). The focus is on "diversified tourism offer (products and services" and "sustainable and responsible tourism management (innovation and quality)."

---

[11] For each of these issues the Action plan 2020 formulates indicative actions and examples of targets by 2020.

To reach these objectives, the countries have to cooperate and to coordinate their actions, not only between the countries, but also between the different ministries and decision-making levels within each country. The pillar coordinator (relevant ministries of the countries) will work closely with cross-border counterparts to develop and implement the Action Plan. The Commission acts as independent facilitator. The European Investment Bank (EIB) contributes significantly to the four pillars (EC 2020: 69–73).

The EUSAIR governance architecture (EC 2020b) consists of three levels. First, the political level represented by the Adriatic-Ionian Council/EUSAIR ministerial meeting. Second, the coordinating level led by the Governing Board (GB), assembling EU institutions and National coordinators[12]). Third, the implementing level represented by the four Thematic Steering Groups (TSG), each covering one of the pillars described above. The TSGs are in charge of implementing the Action Plan, while the GB coordinates the work of the TSGs. Special arrangements are in place under Pillar 2, with two sub-groups for transport and energy, respectively. The chair of the four thematic Steering Groups rotates, by two countries involving one non-EU and one EU Member State per pillar.

The Development of an Emerging MRS Governance Arrangement

The dominant MRSs discourse is to develop an integrated approach of the implementation of cross-border and transboundary planning initiatives situated at the regional sea level. Important storylines of this discourse are establishing transboundary cooperation and coordination, multi-level stakeholder dialogues and establishing land-sea interaction. Core actors of the MRS governance arrangement are EU institutions and national and regional planning authorities. MRSs have a focus on common issues, solutions, and actions of relevance for the regional sea. Within MRSs the

---

[12] "The Governing Board is co-chaired by the country chairing pro tempore the AII. Each participating country is represented by two formally appointed national coordinators, i.e. one senior official from the Ministry of Foreign Affairs and one senior official from the line ministry responsible for coordinating EU funds, as well as two formally appointed pillar coordinators and representatives from the Commission, the European Parliament, the Committee of the Regions, the European Economic and Social Committee, the Permanent Secretariat of the Adriatic-Ionian Initiative and the Managing Authority of the Interreg Adriatic Ionian (ADRION) transnational cooperation program" (Gänzle and Mirtl 2019: 244).

emphasis is on strategic coordination and (cross-sectoral and cross-border) cooperation among policies, institutions, and funding sources, as well as multi-stakeholder dialogue. MRSs are formal policy-making processes subject to the rule of the "Three—Nos."

In the emerging MRS governance arrangements Member States, non-EU states, existing and already institutionalized (national, regional, and sub-national) institutions play a paramount role. For example, the Secretariat of the Council of the Baltic Sea States plays a prominent role in the management of the EUSBSR[13] (Gänzle and Mirtl 2019; 249). Illustrative for the central role of Member States is that in all MRSs "the political level is generally represented by the Ministers of Foreign Affairs and, in some cases, the ministers or authorities in charge of EU funds" (EC 2020c)(7), while the "coordination of the MRS within and between participating countries is done by the national coordinators" (EC 2020c: 7). In the multi-level governance MRS architecture the involved actors (regional, EU, national, and local authorities, intergovernmental and non-governmental bodies) have clear roles and responsibilities in the development and the implementation of the MRS. "The role of the implementing bodies (thematic/priority/policy steering or action groups) has grown noticeably, as these are the drivers of the day-to-day implementation of MRS action plans. The EC acknowledges that the MRS key implementers need financial, political, and administrative support to fulfil their tasks, and that further work is required to appropriately empower them with clear mandates and effective decision-making capacity, while ensuring that they have the resources, technical capacity and the skills needed" (EC 2020c: 7).

"All the MRS are making efforts to involve civil society in work on thematic areas. Implementing bodies are increasingly connected with civil society. Participation by local communities strengthens the bottom-up dimension of the MRS actions. (…) In the Baltic Sea strategy, civil society actors, predominantly from higher education and research institutions, are involved in many projects/activities. However, there is scope for increased involvement of the business community, NGOs, and young people. The EUSAIR stakeholder platform is operational (…)" (pp. 8).

---

[13] The CBSS Secretariat co-coordinates one policy area (Policy Area Secure) and two horizontal actions (Neighbours and Climate).

5.4   LIQUID INSTITUTIONALIZATION, POWER
AND REFLEXIVITY IN TRANSBOUNDARY REGIONALIZATION

How can we understand the process of liquid institutionalization within transboundary regionalization? As discussed, there is a clear difference between TMSP and MRSs; while the former is an informal governance experiment setting the scene for developing institutional rules, MRSs are more formal and embedded in EU policies.

5.4.1   Liquid Institutionalization of TMSP

In the last decade, MSP has become institutionalized as a governance arrangement to deal with conflicting uses at seas. With the acceptance of the MSPD, MSP stabilized as a governance arrangement at the EU level. Dominant storylines of this discourse are sustainable growth of activities, sustainable use of resources, an optimal distribution of maritime space and an ecosystem-based approach (European Union 2014). The "rules of the game" of this EU MSP governance arrangement are that MSs are responsible to carry out MSP, while the European Commission ensures a framework for cooperation between MSs and neighbouring third countries. However, MSs should aim for coherence of management across sea basins, through transboundary cooperation. The implementation of the MSPD by EU MSs is a process of structuration in which each country develops its own form of Maritime Spatial Planning processes and the development of a Maritime Spatial Plan within the context of already existing planning traditions, cultures and rules (Kirkfeldt et al. 2020). The institutional ambiguity of the EU MSP arrangement concerning transboundary cooperation and coordination leaves room for governmental and non-governmental actors to design, to develop and to experiment with TMSP. Initiated by scientists and supported by DG MARE TMSP governance arrangements emerged. With these emerging governance arrangements scientists in cooperation with national authorities, other governmental and non-governmental actors wanted to get insight in the enabling and constraining conditions given the EU MSP institutional setting, and to formulate recommendations and best practices to overcome existing institutional barriers (national planning path dependency, fragmented governance structure at the regional sea levels, etc.) related to transboundary cooperation at the regional sea level. The TMSP projects discussed in this chapter are all experimental projects in which the participating actors learn from each other in a "safe and controlled" learning setting/environment. In this experimental setting, initiated, coordinated, and guided by scientists, governmental authorities from different levels (sub-national, national, regional) and non-governmental actors are willing to look for joint solutions, develop in interaction common problem analyses and best practices to realize joint transboundary planning arrangements. This is only possible if the involved actors/participants are not bound by the formal rules of their national MSP but can negotiate on personal title. This experimental setting, in which key actors are brought together, who could "informally" explore and discuss the enabling and constraining (institutional) conditions of transboundary cooperation and coordination, contributed to the success of TMSP projects.

What type of liquid institutionalization is characteristic for TMSP? All projects are examples and illustrations of exploring the boundaries of possible institutional change if these rules would be implemented in real formal planning processes. This reflects processes of structuration and elaboration. In earlier research (van Tatenhove 2017), I characterized the projects TPEA and MASPNOSE as examples of a combination of unreflectiveness and structural reflectiveness, because the national planning discourses remain leading and respected, while the institutional rules are not challenged. Plan Bothnia is an example of performative reflectiveness because it presents successfully an attempt to implement forms of transboundary cooperation within the rule systems of Finland and Sweden. The BaltSeaPlan shows elements of performative reflectiveness and reflexivity (the subproject "Pomeranian Bight and Arkona Basin"). BaltSeaPlan has resulted in new planning principles (Pan-Baltic thinking, Spatial efficiency and Connectivity thinking (Gee et al. 2013)), two draft transboundary maritime plans and ideas about new roles of existing institutions (conversion). In the "The Pomeranian Bight and Arkona basin" subproject the discursive space was challenged (by stressing the need for permanent cross-border cooperation), while the (national) institutional setting was also challenged. In addition, the Baltic SCOPE project concluded that a discursive change is needed, by proposing a Pan-Baltic Sea Region Approach to MSP (Kull et al. 2017: 101). The project identified seven best practices (such as creating a permanent MSP framework, developing a joint stakeholder involvement strategy, a taskforce to incorporate EBA in MSP and Pan-Baltic planning) to develop new institutions or to convert existing ones. The "Pan Baltic Scope" project emphasized the need for a discursive change by stressing the importance of land-sea interactions (LSI) as an enabler for MSP. By linking land-based planning to maritime planning not only a discursive change is needed (develop LSI Mindfulness, Cedergren et al. 2019: 79), but also new institutions must be developed (displacement) or the convention of existing ones to new ones (conversion).

To conclude, TMSP projects as experimental governance arrangements show characteristics of Type III (Syrupy) and IV liquid institutionalization. National planning discourses are challenged with experiments of transboundary cooperation and coordination. The dominant form of institutional change is conversion. The formal rules of MSP remain the same, but are transformed, integrated, and enacted in new ways, redirecting existing institutions. Although civil servants and politicians of national governments, Regional Sea Conventions and the EU participate in these projects, the outcomes of TMSP projects only indirectly affect national maritime planning processes and the institutional rules concerning transboundary cooperation and coordination. In other words, in the light of existing MSP planning arrangements TMSP is not an example of liquid institutionalization (Types I and II) and the future will learn whether these projects will be the start of processes of structural elaboration. The success of a further institutionalization of TMSP depends on the role researchers can play as programmers of TMSP projects (setting the agenda and translating the wishes of national governments in best practices), but more importantly whether governmental and non-governmental actors are able to play the role of switchers and successfully transpose developed discourses, rules and resources in experimental projects to the formal MSP planning processes.

5.4.2 Liquid Institutionalization of MRSs

Although MRSs are initiated within a clear EU framework supported with the rule of the "Three—Nos" and clear roles and responsibilities for involved governmental authorities and non-state actors, processes of structural elaboration/structuration take place. The question is whether these processes of structural elaboration result in new institutions, and lastly in a process of stabilization. Gänzle (2018: 342) asks himself a comparable question whether non-statutory arrangements in macro-regions eventually led to the solidification, if not "hardening" (or formalization) of "these" soft arrangements. According to Gänzle (2018: 344) MRSs firstly establish a framework which, despite the "Three Nos", tends to institutionalize "consultation patterns, decision-making procedures, administrative roles and behaviour expectations (…)." Secondly, MSRs provide actors the opportunity to mobilize and defend their interests and to forge policies, and alliances and institutions that accommodate them. Thirdly, they encourage the implementation of a set of interconnected policies (encouraging "diagnostic monitoring"). Fourthly, based on feedback the framework is regularly revised and adjusted. In his analysis of the EUSBSR from an experimental governance perspective Gänzle concludes that a lean governance architecture has been established that draws on existing institutions that have been integrated "into the texture" of EUSBSR governance (2018: 348). The analysis of the EUSBSR showed processes of institutional change as conversion, in which transnational institutions are redesigned and solidify, becoming constitutive elements of macro-regions. "(…) the soft spaces of the Baltic Sea Region is hardening (…)" (2018: 348).

From a strategic relational heuristic perspective Melanie Plangger (Plangger 2019) understands the development of a MRS from the interests and strategic action of actors, the contextual factors and meta-governance. When actors perceive the MRS as an opportunity to realize their interests, they will employ *strategic action* (Plangger 2019: 161), based on an understanding of the context by the actor. MRS is a contextual process of cross-border region-building in which multiple actors participate. The created MRS structure is both constraining and enabling for different actors. This "strategic selectivity" (Jessop 2001: 9) emphasized the different opportunities and possibilities of actors within a MRS. Actors develop "meta-governance" activities, because "they are expected to shape not the process as such, but the environment of the process"; while reflecting upon "the context and strive to change the contextual elements to optimize the realization of their strategies" (Plangger 2019: 162).

To conclude MRSs show characteristic of Type II (*Gelatinous*) liquid institutionalization. The institutionalization of MRSs takes place within the EU discourse of "developing an integrated approach of the implementation of cross-border and transboundary planning initiatives situated at the regional sea level." Within this discursive space, actors have the ability to use rules from different institutional settings and to develop new institutions or institutional rules. The MRS governance arrangement is characterized by structural reflectiveness and has some room to change the institutional rules and converse them to new ones (conversion). The EU framework of MRS and the existing national and regional seas institutions form the institutional settings in which governmental and non-governmental actors operate. The role of MSs is paramount, they are the programmers and switchers of the MRSs.

References

Almodovar, Margarida, Demetrio de Armas, Fátima Lopes Alves, Luis Bentes, Catarina Fonseca, Jordi Galofré, Kira Gee, et al. 2014. TPEA Good Practice Guide—Lessons for Cross-Border MSP from Transboundary Planning in the European Atlantic.

Ardron, J., K. Gjerde, S. Pullen, and V. Tilot. 2008. Marine Spatial Planning in the High Seas. *Marine Policy* 32 (5): 832–839. https://doi.org/10.1016/j.marpol.2008.03.018.

Backer, Hermanni, and Manuel Frias. 2013. Planning the Bothnian Sea—Key Findings of the Plan Bothnia Project. https://helcom.fi/wp-content/uploads/2019/08/Planning-the-Bothnian-Sea.pdf.

Backer, H., U. Bergström, C. Fredricsson, R. Fredriksson, M. Frias, J. Hämäläinen, and L.Z. Snowball. 2013. Planning the Bothnian Sea (Report No. 158).

Cedergren, E., K. Gee, M. Kull, S. Eliasen, and A. Morf. 2019. *Lessons, Stories and Ideas on How to Integrate Land-Sea Interactions into MSP.* Stockholm.

Douvere, Fanny. 2008. The Importance of Marine Spatial Planning in Advancing Ecosystem-Based Sea Use Management. *Marine Policy* 32 (5): 762–771. https://doi.org/10.1016/j.marpol.2008.03.021.

Douvere, Fanny, and Charles N. Ehler. 2007. International Workshop on Marine Spatial Planning, UNESCO, Paris, 8–10 November 2006: A Summary. *Marine Policy* 31 (4): 582–583. https://doi.org/10.1016/j.marpol.2007.02.001.

———. 2009. New Perspectives on Sea Use Management: Initial Findings from European Experience with Marine Spatial Planning. *Journal of Environmental Management* 90 (1): 77–88. https://doi.org/10.1016/j.jenvman.2008.07.004.

EC. 2006. Towards a Future Maritime Policy for the Union. A European Vision for the Oceans and Seas. Green Paper. COM(2006)275final, Volume II—ANNEX. Brussels.

———. 2007. An Integrated Maritime Policy for the European Union. Blue Paper COM(2007)575final. Brussels.

———. 2008. Roadmap for Maritime Spatial Planning: Achieving Common Principles in the EU. COM(2008)791final. Brussels.

———. 2014. Communication from the Commission Concerning the European Union Strategy for the Adriatic and Ionian Region. COM(2014) 357 Final. Brussels.

————. 2020a. Action Plan Concerning the the European Union Strategy for the Adriatic and Ionian Region. Commission Staff Working Document. SWD(2020)57 Final. Brussels.

————. 2020b. Commission Staff Working Document Accompanying the Document on the Implementation of EU Marco-Regional Strategies. SWD(2020) 186 Final. Brussels.

————. 2020c. Report on the Implementation of EU Macro-Regional Strategies. COM(2020) 578 Final. Brussels.

Eliasen, Søren Q., Troels J. Hegland, and Jesper Raakjær. 2015. Decentralising: The Implementation of Regionalisation and Co-Management under the Post-2013 Common Fisheries Policy. *Marine Policy* 62 (December): 224–232. https://doi.org/10.1016/j.marpol.2015.09.022.

European Union. 2014. Directive 2014/89/EU of the European Parliament and of the Council of 23 July 2014 Establishing a Framework for Maritime Spatial Planning. *Offical Journal of the European Union* 2014 (April): 135–145.

Faludi, A. 2010. Beyond Lisbon: Soft European Spatial Planning. *DISP* 46 (182): 14–24. https://doi.org/10.1080/02513625.2010.10557098.

Flannery, Wesley, Geraint Ellis, Geraint Ellis, Wesley Flannery, Melissa Nursey-Bray, Jan P.M. van Tatenhove, Christina Kelly, et al. 2016. Exploring the Winners and Losers of Marine Environmental Governance/Marine Spatial Planning: Cui Bono?/'More than Fishy Business': Epistemology, Integration and Conflict in Marine Spatial Planning/Marine Spatial Planning: Power and Scaping/Surely Not All. *Planning Theory and Practice* 17 (1): 121–151. https://doi.org/10.1080/14649357.2015.1131482.

Flannery, Wesley, Hilde Toonen, Stephen Jay, and Joanna Vince. 2020. A Critical Turn in Marine Spatial Planning. In *Maritime Studies*. Springer. https://doi.org/10.1007/s40152-020-00198-8.

Gänzle, Stefan. 2018. 'Experimental Union' and Baltic Sea Cooperation: The Case of the European Union's Strategy for the Baltic Sea Region (EUSBSR). *Regional Studies, Regional Science* 5 (1): 339–352. https://doi.org/10.108 0/21681376.2018.1532315.

Gänzle, Stefan, and Kristine Kern, eds. 2016. *A "Macro-Regional" Europe in the Making. Theoretical Approaches and Empirical Evidence.* Houndmills, Basingstoke, Hampshire: Palgrave Macmillan. https://doi. org/10.1007/978-1-137-50972-7.

Gänzle, Stefan, and Jörg Mirtl. 2019. Experimentalist Governance Beyond European Territorial Cooperation and Cohesion Policy: Macro-Regional Strategies of the European Union (EU) as Emerging 'Regional Institutions'? *Journal of European Integration* 41 (2): 239–256. https://doi.org/10.108 0/07036337.2019.1580277.

Gee, K., A. Kannen, and B. Heinrichs. 2013. BaltSeaPlan Vision 2030: Towards the Sustainable Planning of the Baltic Sea Space.

Gløersen, Erik, Jörg Balsiger, Battistina Cugusi, and Bernard Debarbieux. 2019. The Role of Environmental Issues in the Adoption Processes of European Union Macro-Regional Strategies. *Environmental Science & Policy* 97 (July): 58–66. https://doi.org/10.1016/j.envsci.2019.04.002.

Gómez-Ballesteros, M., C. Cervera-Núñez, M. Campillos-Llanos, A. Quintela, L. Sousa, M. Marques, F.L. Alves, et al. 2021. Transboundary Cooperation and Mechanisms for Maritime Spatial Planning Implementation. SIMNORAT Project. *Marine Policy* 127 (May): 104434. https://doi.org/10.1016/j.marpol.2021.104434.

Healey, Patsy. 2006. Relational Complexity and the Imaginative Power of Strategic Spatial Planning. *European Planning Studies* 14 (4): 525–546.

Hommes, Saskia. 2012. Report on Cross-Border Maritime Spatial Planning in Two Case Studies. Vol. 2012.

INTERACT. 2014. Macro-Regional Strategy—Sea Basin Strategy: What Is What? http://www.interact-eu.net/library/macro_regional_strategy_sea_basin_strategy_what_is_what_/514/8859.

Jay, Stephen, Fátima L. Alves, Cathal O'Mahony, Maria Gomez, Aoibheann Rooney, Margarida Almodovar, Kira Gee, et al. 2016. Transboundary Dimensions of Marine Spatial Planning: Fostering Inter-Jurisdictional Relations and Governance. *Marine Policy* 65 (March): 85–96. https://doi.org/10.1016/j.marpol.2015.12.025.

Jessop, B. 2001. Institutional Re(Turns) and the Strategic—Relational Approach. *Environment and Planning A* 33 (7): 1213–1235. https://doi.org/10.1068/a32183.

Käppeler, B., S. Toben, G. Chmura, S. Walkowicz, N. Nolte, P. Schmidt, and C. Mohn. 2012. Developing a Pilot Maritime Spatial Plan for the Pomeranian Bight and Arkona Basin (BaltSeaplan Report 9).

Kirkfeldt, Trine Skovgaard, Jan P.M. van Tatenhove, Helle Nedergaard Nielsen, and Sanne Vammen Larsen. 2020. An Ocean of Ambiguity in Northern European Marine Spatial Planning Policy Designs. *Marine Policy* 119 (September): 104063. https://doi.org/10.1016/j.marpol.2020.104063.

Kull, M., J. Moodie, A. Giacometti, and A. Morf. 2017. *Lessons Learned: Obstacles and Enablers When Tackling the Challenges of Cross-Border Maritime Spatial Planning—Experiences from Baltic SCOPE*. Stockholm, Espoo and Gothenburg. http://www.balticscope.eu/content/uploads/2015/07/BalticScope_LL_WWW.pdf.

Kull, M., J.R. Moodie, H.L. Thomas, S. Mendez-Roldan, A. Giacometti, A. Morf, and I. Isaksson. 2021. International Good Practices for Facilitating Transboundary Collaboration in Marine Spatial Planning. *Marine Policy* 132 (October): 103492. https://doi.org/10.1016/j.marpol.2019.03.005.

Li, Shenghui, and Stephen Jay. 2020. Transboundary Marine Spatial Planning across Europe: Trends and Priorities in Nearly Two Decades of Project Work.

Marine Policy 118 (August): 104012. https://doi.org/10.1016/j.marpol.2020.104012.

Maes, F. 2008. The International Legal Framework for Marine Spatial Planning. *Marine Policy* 32 (5): 797–810. https://doi.org/10.1016/j.marpol.2008.03.013.

Maier, Nina, and Till Markus. 2013. Dividing the Common Pond: Regionalizing EU Ocean Governance. *Marine Pollution Bulletin* 67 (1–2): 66–74. https://doi.org/10.1016/j.marpolbul.2012.11.042.

Moodie, John R., Michael Kull, Elin Cedergren, Alberto Giacometti, Andrea Morf, Søren Qvist Eliasen, and Lise Schrøder. 2021. Transboundary Marine Spatial Planning in the Baltic Sea Region: Towards a Territorial Governance Approach? *Maritime Studies* 20 (1): 27–41. https://doi.org/10.1007/s40152-020-00211-0.

Nienaber, Birte, and Christian Wille. 2020. Cross-Border Cooperation in Europe: A Relational Perspective. *European Planning Studies* 28 (1): 1–7. https://doi.org/10.1080/09654313.2019.1623971.

Pastoors, Martin, Saskia Hommes, Frank Maes, David Goldsborough, Birgit de Vos, Marian Stuiver, Bas Bolman, Thomas Kirk Sørensen, and Vanessa Stelzenmüller. 2012. Preparatory Action on Maritime Spatial Planning in the North Sea—MASPNOSE Final Report. May 2012: 1–40.

Plangger, M. 2019. Exploring the Role of Territorial Actors in Cross-Border Regions. *Territory, Politics, Governance* 7 (2). https://doi.org/10.1080/21622671.2017.1336938.

Pomeroy, Robert, and Fanny Douvere. 2008. The Engagement of Stakeholders in the Marine Spatial Planning Process. *Marine Policy* 32 (5): 816–822. https://doi.org/10.1016/j.marpol.2008.03.017.

Rochette, Julien, Sebastan Unger, Dorothée, Herr, David Johnson, Takehiro Nakamura, Tim Packeiser, Alexander Proelss, Martin Visbeck, Andrew, Wright, Daniel Cebrian. 2014. The regional approach to the conservation and sustainable use of marine biodiversity in areas beyond national jurisdiction. Marine Policy 49109-117 S0308597X14000438 https://doi.org/10.1016/j.marpol.2014.02.005

Schultz-Zehden, Angela, Nico Nolte, and Kira Gee. 2013. BaltSeaPlan Vision 2030. Towards the Sustainable Planning of the Baltic Sea Space.

Smith, Joanna. 2018. Keynote: MSP 10 Things. Lessons Learned from the Pacific and Indian Oceans. In *PADDLE Summerschool Brest*. Brest.

Soininen, Niko, Tuomas Kuokkanen, and Daud Hassan. 2015. Comparative and Forward-Looking Conclusions on Transboundary MSP. In *Transboundary Marine Spatial Planning and International Law*, 221–226. Taylor and Francis Inc. http://www.scopus.com/inward/record.url?eid=2-s2.0-84942525262&partnerID=tZOtx3y1.

Soma, Katrine, Jan P.M. van Tatenhove, and Judith van Leeuwen. 2015. Marine Governance in a European Context: Regionalization, Integration and Cooperation for Ecosystem-Based Management. *Ocean and Coastal Management* 117: 4–13. https://doi.org/10.1016/j.ocecoaman.2015.03.010.

St Martin, Kevin, and Madeleine Hall-Arber. 2008. The Missing Layer: Geo-Technologies, Communities, and Implications for Marine Spatial Planning. *Marine Policy* 32 (5): 779–786. https://doi.org/10.1016/j.marpol.2008.03.015.

UNESCO. 2009. Marine Spatial Planning. A Step-by-Step Approach toward Ecosystem-Based Management. http://www.unesco-ioc-marinesp.be/msp_guide.

van Tatenhove, Jan P.M. 2016. The Environmental State at Sea. *Environmental Politics* 25 (1): 160–179. https://doi.org/10.1080/09644016.2015.1074386.

———. 2017. Transboundary Marine Spatial Planning: A Reflexive Marine Governance Experiment? *Journal of Environmental Policy and Planning* 19 (6): 783–794. https://doi.org/10.1080/1523908X.2017.1292120.

van Tatenhove, Jan P.M., and Judith van Leeuwen. 2015. Marine Governance of the North Sea: Patterns of Regionalization. In *Governing Europe's Marine Environment. Europeanization of Regional Seas or Regionalization of EU Policies?* ed. M. Gilek and K. Kern, 183–202. Farnhem: Ashgate Publishing Ltd.

Vince, Joanna, Erik Olsen, Wen-Hong Liu, Juan Luis Suárez, Julia de Vivero, Guifang Xue, Magdalena Matczak, Stephen Jay, et al. 2013. International Progress in Marine Spatial Planning. *Ocean Yearbook Online* 27 (1): 171–212. https://doi.org/10.1163/22116001-90000159.

CHAPTER 6

Conclusions and Reflections

Abstract This chapter summarizes the main argument of the book and draws some conclusions. The main concept—liquid institutionalization—gives insight into processes of institutionalization at the level of regional seas and the high seas. The assumption of this book was that the farther away from shores and beaches processes of liquid institutionalization would increase, because not one authority is responsible for the governing of maritime activities and sectors and there is room to negotiate, to invent and to develop new institutional rules. In Chap. 2, four forms of liquid institutionalization are distinguished: frozen, gelatinous, syrupy and liquid. The first conclusion is that although the liquid institutionalizations of Arctic shipping, Deep Seabed Mining and Transboundary Regionalization show different forms of liquid institutionalization, the dominant types are gelatinous and syrupy. Secondly, to understand processes of liquid institutionalization power should be an integral aspect of the analysis. Thereafter, the chapter discusses the analytical value of liquid institutionalization, from a structural perspective and an agency perspective. The chapter ends with formulating future research needs.

Keywords Liquid institutionalization • Blue governance arrangements • Future research needs • Analytical value

© The Author(s), under exclusive license to Springer Nature   113
Switzerland AG 2022
J. P. M. van Tatenhove, *Liquid Institutionalization at Sea*,
https://doi.org/10.1007/978-3-031-09771-3_6

6.1   MAIN ARGUMENTS AND FINDINGS

This book is about the search of how activities and processes at regional seas and the high seas are governed, by understanding blue governance and the (liquid) institutionalization of blue governance arrangements at sea.

Blue governance is the capacity of governmental and non-governmental actors within blue governance arrangements to govern maritime activities and their consequences in anticipatory and reflexive ways. *Institutionalization* is the structuration and stabilization of (blue) governance arrangements in processes of structuration/elaboration and stabilization/maintenance. The institutionalization of governance arrangements should be understood both from a structural and an agency perspective (see Sect. 6.2).

In this book, I developed the concept of *liquid institutionalization* to give insight into the processes of institutionalization at the level of regional seas and the high seas. Governance processes at the regional seas and high seas are characterized by (1) a fragmented institutional setting, consisting of the institutional rules of different governmental layers (UN, Regional Sea Commissions, the EU and nation states) and the institutional dynamic of several regime complexes of maritime sectors and activities (van Tatenhove 2016), (2) situations of high institutional ambiguity, referring to the mismatch of institutional rules from different institutional settings. The assumption of this book is that because of the fragmented institutional setting, situations of high ambiguity and the invisibility of how maritime activities develop beyond the territorial waters of states, there are no clearly defined rules to guide the governing of maritime activities and their ecological, economic and social consequences. Formulated differently; blue governance arrangements are in a permanent state of flux and becoming without a solid stabilization in institutionalized governance arrangements. The idea behind liquid institutionalization is that the further away from shores and beaches the governing of maritime activities and sectors is complicated because not one authority is responsible and there is room to negotiate, to invent and to develop new institutional rules as the cases of Arctic Shipping, Deep Seabed Mining in the *"Clarion Clipperton Zone"* (CCZ) and Transboundary Regionalization show. Therefore, in Chap. 2, the building blocks of liquid institutionalization are presented: institutionalization, a liquid ontology, institutional change and stability, power and reflexivity. Liquid institutionalization is based on a liquid ontology. This liquid ontology expresses the nature of social being at sea, the kind of things that exists at sea, the conditions of existence, the relations of interdependency and the standards that must be met to fully exist at sea. Liquidity refers to the invisibility of activities, the permanent state of flux and becoming and the ambiguity of governing, but also to the uncertainties of knowing the nature of social being at seas and the conditions of governing. Policy-making and governance processes are complicated and so are processes of institutionalization. Liquid institutionalization is defined as an ongoing process of structuration/elaboration in which blue governance arrangements are continuously negotiated, while in a permanent state of "flux and becoming," as a consequence of contexts of high ambiguity and reflexivity, without a solid stabilization in institutionalized blue governance arrangements. In this sense, liquid institutionalization is a form of institutionalized improvisation. However, policy-making and governance processes always know phases of stabilization, which makes the concept of liquid institutionalization an ideal type in the sense developed by Max Weber. An ideal type is "the construction of certain elements of reality in a logically precise conception" (Gerth and Wright Mills 1982: 59). It is "an abstract model that, when used as a standard of comparison, enables us to see aspects of the real world in a clearer, more systematic way" (Johnson 2000). To understand the different forms of liquid institutionalization, I developed a continuum of four types of liquid institutionalization, based on the building blocks structuration/elaboration—stabilization/maintenance, forms of institutional change and forms of reflectiveness/reflexivity. The four types are ideal types, non-existing in the real world but are analytical tools to understand processes of institutionalization at seas.

Chapters 3, 4 and 5 presented and analysed the institutionalization of blue governance arrangements in Arctic Shipping, Deep Seabed Mining and Transboundary Regionalization. It is expected that in the nearby future three *Arctic shipping governance arrangements* will emerge, related to the opening of three possible shipping routes: the Northeast Passage/Northern Sea Route, the Northwest Passage and the Transpolar Sea Route. These three Arctic shipping governance arrangements institutionalize within the dominant discursive space that shipping in the Arctic is a legitimate activity under the conditions and rules of environmental protection and human safety as formulated in the Polar Code. These governance arrangements differ in types of shipping, the actors involved and the availability of resources, related to the geographical area of the route, and the natural circumstances. Chapter 3 concluded that the future liquid institutionalization of Arctic shipping governance arrangements shows characteristics of Type II (Gelatinous) liquid institutionalization. The analysis of *Deep Seabed Mining governance arrangements* in Chap. 4 showed a diffuse picture. The planning of DSM activities and the institutionalization of governance arrangements is both a process of stabilization in which the existing historical power position of pioneer firms and sponsoring states is defended against change, supported by neo-liberal and neo-mercantilist discourses. At the same time, DSM governance shows processes of structuration and institutional change related to the development of the mining code and the development of the Environmental Management Plan for the CCZ. The analysis showed characteristics of Type II (Gelatinous) and Type III (Syrupy) liquid institutionalization of DSM. In other words, the institutionalization of DSM governance arrangements is the result of a complex interplay of stabilization/maintenance and structuration/elaboration, structural reflectiveness, while institutional changes are characterized by layering, drift and displacement. The analysis of *transboundary regionalization governance arrangements* in Chap. 5 showed two very different faces. On the one hand, the informal governance experiments of Transboundary Maritime Spatial Planning projects and on the other hand the EU initiated formal governance processes of the Sea-Basin Strategies and the Macro Regional Strategies. The TMSP projects as experimental governance arrangements show clear characteristics of Type III (Syrupy) and IV liquid institutionalization. Dominant planning discourses are challenged with experiments of transboundary cooperation and coordination and new institutional rules are suggested. However, these experiments do not change existing stabilized MSP governance arrangements. In the light of these existing MSP planning arrangements TMSP is not an example of liquid institutionalization (Types I: Frozen and II: Gelatinous). Depending on the transformative role of governmental and non-governmental actors to successfully transpose developed discourses, rules and resources within experimental projects to the formal MSP planning processes will set into motion a process of Type III (Syrupy) and IV liquid institutionalization in existing TMSP governance arrangements. MRSs show characteristic of Type II liquid institutionalization because these governance arrangements institutionalize within the existing EU discourse of "cross-border and transboundary planning." Within this discursive space actors have some room to change the institutional rules and converse them to new ones.

Based on the analysis of the cases presented in the previous chapters, the following conclusions can be drawn.

1. Although the liquid institutionalization of Arctic shipping, Deep Seabed Mining and Transboundary Regionalization show different forms of institutionalization, the dominant forms are type II (Gelatinous) and type III (Syrupy). Liquid institutionalization is the result of both structuration/elaboration and stabilization/maintenance. The differences are in the type of reflexivity. Arctic shipping is an example of structural reflectiveness; the dominant discursive space "navigation in the Arctic is a legitimate activity under environmental and safety conditions" is not challenged while China and Russia are able to mobilize rules and resources from different settings. Also, in the Deep Seabed Mining governance arrangement structural reflectiveness is dominant. In the DSM governance arrangement two dominant discourses compete: a neo-liberal discourse and the "Common Heritage of Mankind" (CHM) discourse. Although ISA, developing countries and NGOs promote the CHM discourse, activities in the CCZ are structured by the dominant neo-liberal discourse, which gives mining companies and sponsoring states the possibility to reduce the importance of the CHM discourse and reduce a successfully implementation of the rules of the CHM discourse. Transboundary regionalization shows a mixed picture. While the institutionalization of the MRSs is an example of structural reflectiveness, in which the EU discourse of cross-border cooperation and integration is not challenged, the experimental TMSP governance arrangements are an example of performative reflectiveness because the national MSP discourses are challenged while also new planning and design principles are developed.

2. To understand processes of (liquid) institutionalization, power must be an integral aspect of the analysis. In Chap. 2 the multi-layered power conception of Arts and van Tatenhove (Arts and van Tatenhove 2005) is combined with the mechanisms and power holder positions in network-making power of Castells (2009). Power should be understood as the interplay of power processes on three levels, at policy practices (relational power), at the level of blue governance arrangements (dispositional power) and at the level of political modernization (structural power). Programmers and switchers (power holders) connect the different levels and play a key

role in the programming of blue governance arrangements (in terms of coalitions, rules, resources, and discourses) and the switching between coalitions, rules, resources and discourses of different blue governance arrangements, resulting in new governance arrangements. The multi-layered power dynamics and the role of programmers and switchers are clearly illustrated in the cases.

6.2   ANALYTICAL VALUE
OF LIQUID INSTITUTIONALIZATION

What is the analytical value of using the concept of liquid institutionalization? Analytically liquid institutionalization gives insight into processes of institutional change and stability as the interplay of structure (political modernization) and agency (interactions of actors in policy practices). This results in the institutionalization of blue governance arrangements. A blue governance arrangement consists of actors/coalitions and rules, resources and discourses (as structural properties). These four dimensions change due to structural transformations (political modernization) or interactions in policy practices. To understand these processes of institutionalization of blue governance arrangements at sea, additional analytical building blocks have been introduced, such as liquid ontology, gradual institutional change, reflexivity and power (see Fig. 6.1).

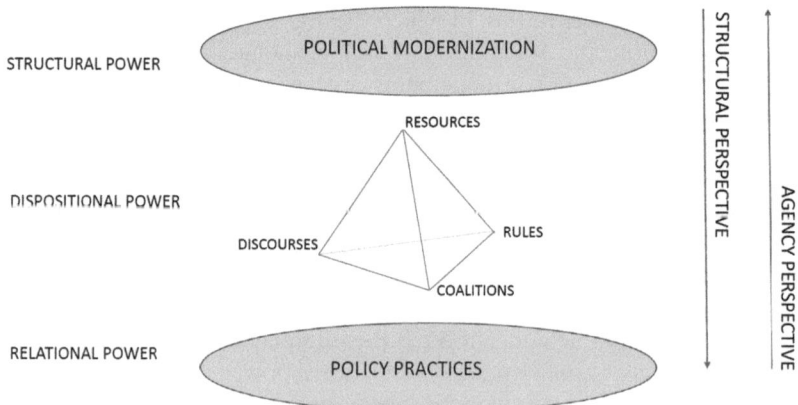

Fig. 6.1 The interplay of political modernization, policy practices and governance arrangements

In earlier work (Arts and van Tatenhove 2006), we made a distinction between an institutional analysis and an analysis of strategic conduct. The former analysed policy-making and governance from a structural perspective, by taking processes of political modernization as starting point of analysis and the way these affect the structural properties of policy arrangements. The latter has as starting point of analysis the actions, ideas, and influence of actors in day-to-day interactions, and how in interactions rules, resources and discourses are used in order to get things done. By doing so the structural properties of a governance arrangements are modified or reproduced. Instead of institutional analysis and analysis of strategic conduct I made the distinction between structural perspective and an agency perspective.

Analysed from a *structural perspective*, blue governance arrangements develop within specific institutional contexts or settings at sea (as a result of processes of political modernization). Processes of Political Modernization enable and constrain the negotiations and deliberations between actors in interactions, because structural processes of societal and political transformations affect the unequal division and availability of resources, the discourses (or discursive spaces) which are perceived as legitimate by actors, the legitimate rules and procedures of the game and possible coalitions of actors which could be formed in interactions. This perspective starts with how structural power results in the positioning of actors in a blue governance arrangement, the unequal allocation of resources and the ability of actors to use available resources. Structural and dispositional power processes also give insight in the role of programmers and switchers.

From an *agency perspective* the analysis focuses on the interaction between actors and coalitions within specific policy practices (of a blue governance arrangement); for example, negotiations about an exploration permit for a firm to explore an area for deep seabed mining by ISA, or the negotiations between governmental and non-governmental about the implementation of Action Plans in Macro Regional Strategies. In these interactions, actors and coalitions have access to and the ability to mobilize specific resources, while their negotiations are guided by existing rules of the game in the context of discursive spaces (discourses). At the same time, during these interactions new rules could be negotiated, competing discourses emerge and new resources become available. This change in structural properties could result in a transformation of the governance arrangement. In the agency perspective, the power analysis starts with relational and dispositional power, that is, how actors use resources to do thing otherwise and the way they are positioned *vis-à-vis* each other in a governance arrangement, but also the strategic role programmers and switchers play in the programming and connecting of governance arrangements.

Following Mahoney and Thelen (2010) different forms of gradual institutional change can be distinguished within a governance arrangement, but next to these institutional changes in governance arrangements, broader processes of institutional change such as structural elaboration could take place in which the structural process of political modernization (the structural conditions) changes. Changes in individual governance arrangements will only indirectly contribute to structural elaboration and processes of political modernization, which are only indirectly related to (human) agency. The unintended outcomes of interactions and outcomes of many different (blue) governance arrangements could result in changes in the structural conditions (processes of political modernization).

6.3   FUTURE RESEARCH NEEDS

This book presented liquid institutionalization as an analytical framework to understand, describe and analyse types of liquid institutionalization of different blue governance arrangements at the regional seas and high seas. The case analyses show the different dynamics of maritime activities, and the framework makes a comparative analysis possible to understand the similarities and differences of the institutionalization of different blue governance arrangements.

Future research should focus on the prescriptive value of the framework, by formulating design principles and different forms of (institutional) capacity building related to the different types of liquid institutionalization. In addition, a focus on the institutional capacity building of actors, which refers to the capacity of actors in blue governance arrangements within a specific institutional setting to mobilize knowledge and relational resources, in order to change the rules of the game would be beneficial. "Improving the institutional capacity of governance arrangements refers to the ability of coalitions to redefine and change the rules of the game by using discourses, and mobilizing knowledge and relational resources in order to change the pathways of policy-making" (van Tatenhove 2015). The assumption is that institutional capacity building differs for the different types of liquid institutionalization and stages of power. Finally, further research should identify the enabling and constraining conditions of institutional capacity building for each type of liquid institutionalization and the consequences for the performance of marine policies.

REFERENCES

Arts, Bas, and Jan P.M. van Tatenhove. 2005. Policy and Power: A Conceptual Framework between the 'Old' and 'New' Policy Idioms. *Policy Sciences* 37 (3–4): 339–356. https://doi.org/10.1007/s11077-005-0156-9.

———. 2006. Political Modernisation. In *Institutional Dynamics in Environmental Governance*, 21–43. Dordrecht: Springer Netherlands. https://doi.org/10.1007/1-4020-5079-8_2.

Castells, M. 2009. *Communication Power*. Oxford: Oxford University Press.

Gerth, H.H., and C. Wright Mills, eds. 1982. *From Max Weber: Essays in Sociology*. London, Boston, Melbourne and Henley: Routledge & Kegan Paul. Reprint.

Johnson, Allan G. 2000. *The Blackwell Dictionary of Sociology: A User's Guide to Sociological Language*. John Wiley & Sons Incorporated Creation.

Mahoney, James, and Kathleen Thelen. 2010. A Theory of Gradual Institutional Change. *Explaining Institutional Change: Ambiguity, Agency, and Power*. 1–37. https://doi.org/9780521134323.

van Tatenhove, Jan P.M. 2015. Marine Governance: Institutional Capacity-Building in a Multi-Level Governance Setting. In *Governing Europe's Marine Environment. Europeanization of Regional Seas or Regionalization of EU Policies?* 35–52. Ashgate.

———. 2016. The Environmental State at Sea. *Environmental Politics* 25 (1): 160–179. https://doi.org/10.1080/09644016.2015.1074386.

Index[1]

---

[1] Note: Page numbers followed by 'n' refer to notes.

© The Author(s), under exclusive license to Springer Nature Switzerland AG 2022

J. P. M. van Tatenhove, *Liquid Institutionalization at Sea*,
https://doi.org/10.1007/978-3-031-09771-3